# CELLULAR AND MOLECULAR MECHANISMS IN HYPERTENSION

# ADVANCES IN EXPERIMENTAL MEDICINE AND BIOLOGY

## Recent Volumes in this Series

A Continuation Order Plan is available for this series. A continuation order will bring delivery of each new volume immediately upon publication. Volumes are billed only upon actual shipment. For further information please contact the publisher.

# CELLULAR AND MOLECULAR MECHANISMS IN HYPERTENSION

Edited by

## Robert H. Cox

Bockus Research Institute
The Graduate Hospital
Philadelphia, Pennsylvania

PLENUM PRESS • NEW YORK AND LONDON

Library of Congress Cataloging-in-Publication Data

Research Symposium on Cellular and Molecular Mechanisms of
  Hypertension: Implications for Pathogenesis and Treatment (1989 :
  Philadelphia, Pa.)
    Cellular and molecular mechanisms in hypertension / edited by
  Robert H. Cox.
        p.   cm. -- (Advances in experimental medicine and biology ; v.
  308)
    "Proceedings of the Fifth Annual Research Symposium on Cellular
  and Molecular Mechanisms of Hypertension: Implications for
  Pathogenesis and Treatment, held November 2-3, 1989, in
  Philadelphia, Pennsylvania"--T.p. verso.
    Includes bibliographical references and index.
    ISBN 0-306-44084-9
    1. Hypertension--Pathophysiology--Congresses.  2. Hypertension-
  -Molecular aspects--Congresses.  3. Myocardium--Pathophysiology-
  -Congresses.  4. Vascular smooth muscle--Pathophysiology-
  -Congresses.   I. Cox, Robert H.  II. Title.  III. Series.
    [DNLM: 1. Hypertension--physiopathology--congresses.
  2. Hypertension--therapy--congresses.   W1 AD559 v. 308 / WG 340
  R432c 1989]
  RC685.H8R42   1989
  616.1'3207--dc20
  DNLM/DLC
  for Library of Congress                              91-39394
                                                           CIP

Proceedings of the Fifth Annual Research Symposium on Cellular and Molecular
Mechanisms of Hypertension: Implications for Pathogenesis and Treatment,
held November 2-3, 1989, in Philadelphia, Pennsylvania

ISBN 0-306-44084-9

© 1991 Plenum Press, New York
A Division of Plenum Publishing Corporation
233 Spring Street, New York, N.Y. 10013

Printed in the United States of America

# PREFACE

Hypertension is recognized to be one of the major risk factors for the development of peripheral vascular disease. The last decade has witnessed several major advances in therapy for hypertension, including the development of angiotensin-converting enzyme inhibitors and calcium channel blockers. These compounds have greatly improved the ability to control blood pressure and to reduce the impact of this risk factor on morbidity and mortality. In spite of these advances, cardiovascular disease remains a major health problem in most modern industrialized countries with related deaths exceeding those from all other causes combined. In contrast to these advances in therapy, our understanding of the basic mechanisms responsible for the pathogenesis of hypertension remains incomplete.

Recent studies have produced new insights into the nature of the regulation of muscle contraction in both heart and blood vessels as well as the changes in muscle function that occur in hypertension. However, the effects of antihypertensive therapy, both in terms of restoring normal function and in producing reversal of hypertension-associated changes, has not been as thoroughly studied, especially in the vasculature. Studies in the heart suggest that the efficacy of different therapeutic agents in restoring normal function and reversing hypertensive changes vary substantially with the mechanism of action of the therapeutic agent. It has also been recently determined that some therapeutic agents produce adverse effects on plasma lipid profiles, which could lead to the secondary acceleration of the atherosclerotic process, while at the same time normalizing blood pressure.

The purpose of this symposium was to bring together a group of experts to discuss current concepts and research in the area of the cellular and molecular biology of cardiac and vascular smooth muscle. The presentations were planned to introduce the topics and tools associated with the study of the intrinsic properties of muscle, both at the single molecule (molecular) level and at the membrane (cellular) level. Presentations were organized to discuss normal function, alterations in function associated with hypertension, and the effects of antihypertensive therapy on function.

The symposium was divided into vascular and cardiac sections, each of which comprised one day of the meeting. Two sessions were organized for each day: one emphasized membrane events involved in the regulation of cell function, and the other intracellular and molecular mechanisms. In addition, each day had a featured lecture by a recognized expert whose role was to synthesize the information presented by the other speakers to provide an overview and summary of concepts related to cellular and molecular function, changes associated with hypertension and the role of antihypertensive therapy. The two individuals invited as featured lecturers are internationally recognized experts in the area of vascular and cardiac changes associated with hypertension, and were well qualified to perform this task.

This symposium was the Fifth Annual Research Symposium sponsored by The Graduate Hospital. It was particularly important because it coincided with the 100th anniversary of the operation of The Graduate Hospital at the location of 19th and Lombard Streets. In addition, this year was the 30th anniversary of the founding of the Bockus International Society and the Bockus Research Institute which honor the contributions of Dr. Henry L. Bockus to our institution. As such, this symposium took on additional meaning to The Graduate Hospital, its house and attending staff, and administration. We are all honored to have been a part of this important and landmark activity.

# CONTENTS

## VASCULAR SMOOTH MUSCLE

### MEMBRANE MECHANISMS

### INTRACELLULAR MECHANISMS

## CARDIAC MUSCLE

### MEMBRANE MECHANISMS

### INTRACELLULAR MECHANISMS

# IONIC CHANNELS OF VASCULAR SMOOTH MUSCLE

# IN HYPERTENSION

Nancy J. Rusch and William J. Stekiel

Department of Physiology
Medical College of Wisconsin
Milwaukee, WI 53226

Altered handling of ions by the vascular muscle membrane is viewed as a potential mechanism contributing to the development and maintenance of high blood pressure. This hypothesis is based on the premise that disturbances in membrane function may enhance vascular reactivity and increase total peripheral resistance. However, while there is a large body of evidence linking an increase in total peripheral resistance with the established stage of hypertension in many animal models of hypertension and in human essential hypertension (for review, 1), pinpointing cellular membrane abnormalities related to the genesis or maintenance of hypertension has been difficult.

In particular, the study of ionic flux through voltage-dependent membrane channels has been hampered by the lack of a method for isolating and directly measuring transmembrane ionic movement. It is only recently that the patch-clamp technique (2) has enabled us to detect with molecular resolution the presence and characteristics of different membrane ionic currents. This method provides a powerful tool to compare directly the properties of membrane ionic channels in vascular smooth muscle from normal and hypertensive animals. Using this technique, our studies initially focused on comparing voltage-dependent calcium channel currents in vascular muscle cells from Wistar Kyoto (WKY) and spontaneously hypertensive (SHR) rats. The SHR was a logical model of hypertension to explore since it manifests an increased total vascular resistance (3), and its vascular muscle displays altered membrane handling of calcium (4).

In an initial study, we compared whole-cell calcium currents between azygos venous cells from 1-3 day old neonatal WKY and SHR (5). The peak amplitudes of the two calcium current types in vascular muscle, termed transient (T-type) current and long-lasting (L-type) current, were measured separately. This was possible since the T-type current clearly activated and inactivated at more negative membrane potentials than the L-type current in this preparation (5,6). Figure 1 shows that by depolarizing from either a holding potential of -80 mV or -30 mV to elicit T-type or L-type current, respectively, venous cells from WKY predominantly exhibited T-type current whereas similar vascular cells from SHR predominantly exhibited L-type current. Since the azygos venous cells were obtained from neonatal SHR before the development of functional adrenergic innervation (7) or significant hypertension (8), it was deemed unlikely that neurogenic factors or a higher perfusion pressure could have influenced the vascular membrane. Rather, we proposed that the altered proportions of T-type and L-type channels activated by depolarization in the SHR were likely genetically determined, and theorized that the larger component of sustained L-type calcium current in the SHR venous cells might be related to the reduced venous compliance implicated in hypertension (for review, 9).

In keeping with the concept that hypertension in the SHR may represent a universal membrane defect (10), we have begun to characterize membrane calcium channels in mesenteric arterial muscle cells from WKY and SHR. Mulvany *et al* reported that mesenteric arterial cells from 4 week old SHR show an increased sensitivity to calcium in the presence of norepinephrine, suggesting that altered handling of calcium by the SHR arterial membrane may be genetically determined (4). More recently, small amplitude T-type and L-type calcium currents have been described in mesenteric arterial cells from adult WKY (11). In these experiments, high external barium concentrations (115 mM) were used to increase the amplitude of the calcium channel current. However, since increasing external calcium and barium concentrations result in a rightward shift of the current-voltage relationship for calcium channel activation (12), we currently are using more physiological cation concentrations (2-10 mM calcium or barium) to better define the membrane potentials at which calcium channel current might activate and inactivate *in situ*.

Figure 2 demonstrates that depolarization from different holding potentials in 10 mM barium activates only L-type current in most mesenteric arterial cells from adult WKY and SHR. This high-threshold calcium channel current shows threshold activation between -20 mV and 0

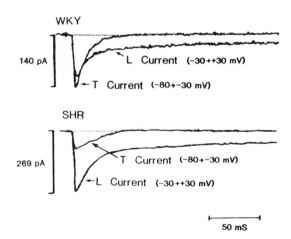

Figure 1: Whole-cell calcium currents in single WKY (upper tracings) and SHR (lower tracings) azygos venous cells. The WKY cell showed predominantly T-type current, while the SHR cell showed more L-type current. T-type current was calculated by digital subtraction of current obtained during a test pulse from -30 mV to 0 mV from that obtained during a test pulse from -80 mV to 0 mV. 20 mM calcium was used in the external solution. (From ref 5 by permission of the American Heart Association, Inc.)

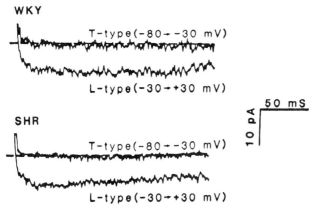

**WKY**

T-type(-80→-30 mV)

L-type(-30→+30 mV)

**SHR**

T-type(-80→-30 mV)

L-type(-30→+30 mV)

50 mS

10 pA

Figure 2: Depolarizing pulses from holding potentials of -80 mV or -30 mV in mesenteric arterial cells from WKY (upper tracings) and SHR (lower tracings). Cells from both rat strains show mainly L-type current. 10 mM barium was used in the external solution.

mV and maximally activates at +30 mV to +40 mV, regardless of the rat strain from which the cells originate. This finding concurs with our previous findings in WKY and SHR azygos venous cells, where we found no evidence to suggest that the current-voltage relationship *per se* of T-type or L-type channels was altered (5). However, the small amplitude of the macroscopic calcium current in rat mesenteric arterial cells will necessitate the measurement of calcium currents in large sample populations of WKY and SHR cells before a conclusive statement can be made regarding calcium current types or their relative amplitudes. In this respect, comparing calcium currents between arterial muscle from normotensive rats and different rat models of hypertension will be challenging. Vascular muscle cells from rat aorta and caudal artery also show low amplitude calcium currents (13,14), which will make quantitative comparisons more difficult.

Since individual types of voltage-dependent ionic channels activate and inactivate in different ranges of membrane potential, the initial holding potential of the cell and the amplitude of the depolarizing pulse will determine the type of ionic current that will be activated, as well as its amplitude. This principle has been applied extensively to the analyses of ionic channel activation and inactivation in voltage-clamped cells. However, it is becoming increasingly clear that the voltages used for holding potentials in standard voltage-clamp experiments often are not in the physiological range of membrane potential. This has led our two laboratories to consider how membrane potential and calcium channel function may be linked *in situ* in vascular muscle cells from WKY and SHR.

Until recently, most membrane potential measurements by the intracellular microelectrode method were performed in pinned *in vitro* vascular segments. Using this technique, vascular resting membrane potentials usually were found in the voltage range of -50 mV to -60 mV (15-19). Since the negative range of these cellular "holding potentials" is below the threshold voltage for calcium channel activation and inactivation, it was suggested that depolarization of intact vascular smooth muscle from resting membrane potential would sequentially activate low-threshold T-

3

type channels and high-threshold L-type channels (5,11). However, recent measurements of membrane potential in rat arteries and veins *in situ* (19,20) show that this concept cannot be universally applied, especially in models of hypertension. Figure 3 shows recordings of intracellular membrane potentials in *in situ* mesenteric arteries of anesthetized WKY and SHR. Both WKY (-45 mV) and SHR (-38 mV) *in situ* membrane potentials are significantly smaller in magnitude when compared to membrane potentials of -60 mV measured in similar pinned *in vitro* vascular segments from both strains. Such depolarization is likely the result of efferent sympathetic neural input, since denervation with 6-hydroxydopamine attenuates the depolarization. The more depolarized mesenteric arterial muscle in SHR relative to WKY can be attributed to the higher level of efferent sympathetic neural regulation in the SHR (1).

The depolarization of these muscle cells *in situ* would bring "resting" membrane potential closer to the threshold for L-type calcium channel activation, suggesting that membrane potentials in intact vascular muscle cells may be more closely coupled to activation of L-type calcium channels than suggested previously. This is demonstrated in Figure 4 which shows that at the average *in vitro* membrane potential of -58 mV in WKY or SHR mesenteric arteries, depolarizing pulses of more than 40 mV must be applied to activate L-type calcium current. However, calcium current is activated by smaller pulse amplitudes at a holding potential equal to the *in situ* WKY resting membrane potential of -43 mV. In addition, from a holding potential equal to the SHR *in situ* membrane potential of -38 mV, the amplitude of the L-type current is further enhanced. These voltage-clamp measurements were made with 10 mM barium in the external solution to enhance the small calcium channel currents. Thus, the coupling of membrane potential with L-type current activation is likely even tighter *in situ* at physiological concentrations of external calcium, where the current-voltage relationship for calcium channel activation presumably would be shifted toward more negative potentials. T-type calcium channels, in contrast, would be partially inactivated at *in situ* membrane potentials, especially at the more depolarized potential in the SHR. This suggests a minor role for this channel type in the maintenance of hypertension by the mesenteric circulation in the SHR model.

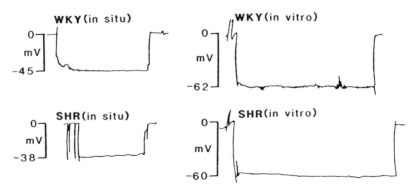

Figure 3: Intracellular recording of membrane potential in WKY and SHR mesenteric arterial cells. Measurements were made in cells from intact arteries in anesthetized rats (*in situ*) or in pinned segments of similar vascular tissue (*in vitro*).

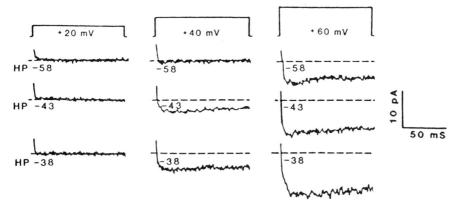

Figure 4: Activation of calcium current from different holding potentials in a SHR mesenteric arterial cell. The holding potentials represent respectivly: the *in vitro* membrane potential of -58 mV, the WKY *in situ* membrane potential of -43 mV, and the SHR *in situ* membrane potential of -38 mV. Since WKY and SHR calcium currents show similar voltage-dependent activation, similar results are obtained in WKY cells.

Intact WKY rat tail arteries also show the presence of a smaller membrane potential *in situ* (19). Compared to a membrane potential of -55 mV in *in vitro* caudal arteries, the average value of *in situ* membrane potentials was -38 mV when mean blood pressure was 100 mmHg. However, when mean arterial pressure spontaneously increased to 140 mmHg in these rats, the average *in situ* membrane potential value was reduced to -31 mV. It is interesting to compare this value with recent steady-state inactivation data in voltage-clamped caudal arterial cells from normotensive Sprague-Dawley rats. These cells show almost complete inactivation of the T-type calcium channel at -31 mV, whereas the majority of L-type channels are available for activation at this membrane potential (14). This again suggests that in some arteries, the T-type channel likely plays at best a minor role in the activation of arterial muscle in hypertension. In contrast, the hypertension-induced depolarization of arterial muscle cells may enhance electrical coupling between the vascular muscle membrane and the L-type calcium channel.

The results of these studies suggest that multiple electrophysiological variables may be altered in vascular smooth muscle cells from SHR. Evidence suggests that voltage-dependent membrane channels are altered genetically in the venous muscle membrane from neonatal SHR. Further studies are being conducted to determine if these findings can be extrapolated to the arterial side of the circulation in the adult SHR. In addition, experiments measuring *in situ* membrane potentials in WKY and SHR vascular muscle suggest that hypertension results in depolarization of arterial smooth muscle. Since activation of voltage-dependent calcium channels is linked intrinsically to the membrane potential of the vascular muscle cell, we are investigating how changes in vascular membrane potential in hypertension may alter calcium channel function.

## ACKNOWLEDGMENT

Supported by grants HL-40474 and HL-29587 from the National Institutes of Health.

# REFERENCES

1. Stekiel WJ. Electrophysiological mechanisms of force development by vascular smooth muscle membrane in hypertension. In: *Blood Vessel Changes in Hypertension: Structure and Function*, Vol. II, R.M.K.W. Lee (ed), Boca Raton: CRC Press, Boca Raton, 1989, p 127.

2. Hamill OP, Marty A, Naher S, Sakmann B, Sigworth FJ. Improved patch-clamp techniques for high-resolution current recording from cells and cell-free membrane patches. *Pflügers Arch* 39: 85, 1981.

3. Frohlich ED. Is the spontaneously hypertensive rat a model for human hypertension? *J Hypertension* 4: S15. 1986.

4. Mulvany MJ, Korsgaard N, Nyborg M. Evidence that the increased calcium sensitivity of resistance vessels in spontaneously hypertensive rats is an intrinsic defect of their vascular smooth muscles. *Clin Exp Hypertens* 3: 749, 1981.

5. Rusch NJ, Hermsmeyer K. Calcium currents are altered in the vascular muscle cell membrane of spontaneously hypertensive rats. *Circ Res* 63: 997, 1988.

6. Sturek M, Hermsmeyer K. Calcium and sodium channels in spontaneously contracting vascular muscle cells. *Science* 233: 475, 1986.

7. Ljung B, Stage D, Carlsson C. Postnatal ontogenetic development of neurogenic and myogenic control in the rat portal vein. *Acta Physiol Scand* 94: 112, 1975.

8. Lais LT, Rios LL, Boutelle S, DiBona GF, Brody MJ. Arterial pressure development in neonatal and young spontaneously hypertensive rats. *Blood Vessels* 14: 277, 1977.

9. Safar ME, London GM. Venous system in essential hypertension, *Clin Sci* 69: 497, 1985.

10. Kwan CY. Dysfunction of calcium handling by smooth muscle in hypertension. *Can J Physiol Pharmacol* 63: 366, 1985.

11. Bean BP, Sturek M, Puga A, Hermsmeyer K. Calcium channels in muscle cells isolated from rat mesenteric arteries: Modulation by dihydropyridine drugs. *Circ Res* 59: 229, 1986.

12. Rorsman P, Trube G. Calcium and delayed potassium currents in mouse pancreatic B-cells under voltage-clamp conditions. *J Physiol* 374: 531, 1986.

13. Toro L, Stefani E. $Ca^{2+}$ and $K^+$ current in cultured vascular smooth muscle cells from rat aorta. Pflügers Arch 408: 417, 1987.

14. Wang R, Karpinski E, Pang PKT. Two types of calcium channels in isolated smooth muscle cells from rat tail artery. *Am J Physiol* 256: H1361, 1989.

15. Hermsmeyer K. Electrogenesis of increased norepinephrine sensitivity of arterial vascular muscle in hypertension. *Circ Res* 38: 362, 1976.

16. Droogmans G, Raeymaekers L, Casteels R. Electro- and pharmacomechanical coupling in the smooth muscle cells of the rabbit ear artery. *J Gen Physiol* 70: 129, 1977.

17. Haeusler G. Relationship between noradrenaline-induced depolarization and contraction in vascular smooth muscle. *Blood Vessels* 15: 46, 1978.

18. Harder DR, Belarinelli L, Sperelakis N, Rubio R, Berne RM. Differential effects of adenosine and nitroglycerin on the action potentials of large and small coronary arteries. *Circ Res* 44: 176, 1979.

19. Bryant HJ, Harder DR, Pamnani MB, Haddy FJ. In vivo membrane potentials of smooth muscle cells in the caudal artery of the rat. *Am J Physiol* 249: C78, 1985.
20. Stekiel WJ, Contney SJ, Lombard JH. Small vessel membrane potential, sympathetic input, and electrogenic pump rate in SHR. *Am J Physiol* 19: C547, 1986.

# REGULATION OF THE $Ca^{2+}$ SENSITIVITY OF

# VASCULAR SMOOTH MUSCLE CONTRACTILE ELEMENTS

Junji Nishimura and Cornelis van Breemen

Department of Pharmacology
University of Miami, School of Medicine
Miami, FL 33101

## INTRODUCTION

$Ca^{2+}$ is assumed to be the primary regulator of vascular smooth muscle contractility (1). In addition, receptor stimulation may modulate the $Ca^{2+}$ sensitivity of vascular smooth muscle myofilaments, probably due to activation of protein kinase C (PKC). Morgan and Morgan (2,3) were the first to measure tension simultaneously with intracellular $Ca^{2+}$ concentration ($[Ca^{2+}]_i$) in strips of ferret portal vein, using the photoprotein aequorin. They found that $\alpha$-adrenergic activation induced a peak of light emission during the period of force development which fell close to the basal value during force maintenance. It has also been reported that phorbol esters which activate PKC (4), induce contraction in intact vascular smooth muscle (5-9), and shift the pCa-tension curve to the left in permeabilized smooth muscle (10,11). Although these reports tend to support a role for PKC in enhancing $Ca^{2+}$ sensitivity of vascular smooth muscle myofilaments, they fail to establish a clear link between receptors, G proteins, PKC and the myofilaments.

While activation of PKC is thought to enhance smooth muscle force development, cAMP and cGMP induce relaxation through activation of cAMP dependent protein kinase (PKA) and cGMP dependent protein kinase (PKG). There has been much discussion about whether cyclic nucleotides modulate smooth muscle contraction directly through an effect on the contractile elements or only indirectly through changes in transmembrane $Ca^{2+}$ fluxes. Adelstein *et al.*(12) initially proposed a model whereby cAMP could induce smooth muscle relaxation by inhibiting myosin light chain (MLC) phosphorylation via phosphorylation of myosin light chain kinase (MLCK) by PKA. Unfortunately a number of experiments using either intact or detergent skinned smooth muscle failed to substantiate a physiologic role for this proposed mechanism, leading to the suggestion that second messenger-induced modulations of smooth muscle contraction operate indirectly through their effect on membrane $Ca^{2+}$ transport mechanisms (13,14,15). However, our recent use of $\alpha$-toxin permeabilization for the study of modulation of smooth muscle calcium sensitivity has verified its up-regulation by PKC and down-regulation by cyclic nucleotides, as described below.

*Cellular and Molecular Mechanisms in Hypertension*
Edited by R.H. Cox, Plenum Press, New York, 1991

This α-toxin treated muscle model also proved to be helpful in exploring the mechanisms underlying the regulation of Ca²⁺ sensitivity and perhaps the regulation of crossbridge cycling rate. We found that merely changing the substrate conditions, namely changing the concentration of ATP and/or creatine phosphate (CP), can cause dramatic changes of the Ca²⁺ sensitivity of the α-toxin permeabilized rabbit mesenteric artery. The data showed that increasing the population of the actomyosin-ADP (AM-ADP) complex by accumulation of ADP inside the fiber or inhibition of ADP release from the crossbridge may be responsible for increasing the Ca²⁺ sensitivity. The possible role of caldesmon in the latter process will also be discussed.

## MATERIALS AND METHODS

The α-toxin permeabilized arterial preparations have been described before (11,16). Small rings (about 250 µm diameter and about 500 µm long) from the first branch of superior mesenteric artery from male Wistar Kyoto rats or third branch of the superior mesenteric artery from New Zealand white rabbits were prepared under binocular microscope and two tungsten wires (40 µm diameter, California Fine Wire Company) were passed through the lumen. One wire was fixed to the chamber and the other was attached to a force transducer (U gage, Sinko Co. Ltd, Tokyo, Japan). Alternatively, small strings of circular muscle were dissected free and mounted by fine silk threads. Permeabilization was accomplished by incubating the arterial segments with Staphylococcal α-toxin (30 µg protein/ml; α-toxin was a generous gift from Dr. R.J. Hohman, NIH) for 15 minutes in the cytoplasmic substitution solution (CSS). The experimental CSS solution contained 2 mM EGTA, 130 mM K propionate, 3.1-4.0 mM MgCl₂, 2-4 mM Na₂ATP, 10-20 mM CP, 20 mM Tris maleate (pH = 6.8), and indicated concentrations of free Ca²⁺. The apparent binding constant used for the Ca²⁺-EGTA complex was $10^6$/M. The concentrations of MgCl₂ (or Mg²⁺), ATP and CP in the CSS solution were changed as indicated in the figures or figure legends. Experiments were carried out at 37°C except for those in Figure 1, which were carried out at room temperature.

## RESULTS

Figure 1A,B shows that α-toxin effectively permeabilizes the arterial smooth muscle cells. After α-toxin treatment the fibers developed tension when exposed to Ca²⁺ concentrations encountered in the cytoplasm of activated intact cells. The observation that these Ca²⁺ induced contractions were at least as big as contractions of the intact artery in response to maximally effective concentrations of norepinephrine (NE) and K assures that most if not all the fibers were adequately permeabilized to ions and high energy phosphates (16,17). In this α-toxin permeabilized rabbit mesenteric artery bathed in $10^{-7}$ M Ca²⁺ solution buffered by 2 mM EGTA, NE ($10^{-4}$ M) caused a sustained contraction (Fig. 1C) that varied in amplitude from one tissue to another (34 ± 35% of the maximal Ca²⁺ induced force; mean ± SD, n = 7). Inclusion of $10^{-4}$ M guanosine-5'-O-(2-thiodiphosphate) (GDP-β-S) in the bathing solution almost completely abolished the NE-induced contractions (1D; 6 ± 4%, n = 4). Conversely, addition of $10^{-4}$ M GTP to the bathing solution enhanced the NE induced contractions to 65 ± 15% of maximum Ca²⁺-induced force (1E; n = 4).

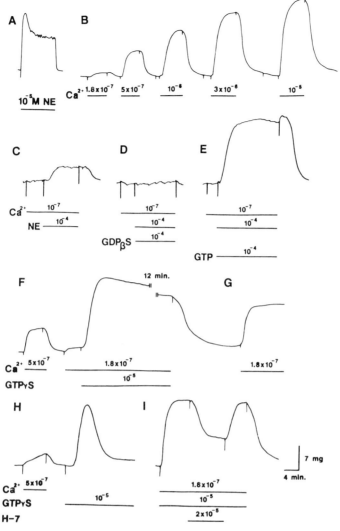

Figure 1: (A) The intact arterial contraction induced by $10^{-5}$ M NE in normal PSS. (B) Tension development induced by different $Ca^{2+}$ concentrations. After the $\alpha$-toxin treatment the indicated concentrations of $Ca^{2+}$ were applied to the permeabilized fiber. After the tension reached steady state, the fiber was relaxed by the 4 mM EGTA 0 M $Ca^{2+}$ solution. (C) $10^{-4}$ M NE-induced contraction in $10^{-7}$ M $Ca^{2+}$ solution after the SR was depleted by 25 mM caffeine. (D) Effect of $10^{-4}$ M GDP-$\beta$-S on $10^{-4}$ M NE-induced contraction. (E) Effect of $10^{-4}$ M GTP on $10^{-4}$ M NE-induced contraction. (F) First the SR was loaded with a 5 x $10^{-7}$ M $Ca^{2+}$ solution, then, 1.8 x $10^{-7}$ M $Ca^{2+}$ solution was applied. After the contraction induced by 1.8 x $10^{-7}$ M $Ca^{2+}$ reached a steady state $10^{-5}$ M GTP-$\gamma$-S was added. GTP-$\gamma$-S caused a large phasic contraction followed by greatly elevated maintained tone. (G) After the GTP-$\gamma$-S was washed out and the fiber was relaxed by 4 mM EGTA 0 M $Ca^{2+}$ solution, 1.8 x $10^{-7}$ M $Ca^{2+}$ solution (without GTP-$\beta$-S) was re-added. Contraction induced by this procedure became much bigger than that before the exposure to GTP-$\gamma$-S and reached almost the same level of the sustained part of the contraction induced by $10^{-5}$ M GTP-$\gamma$-S. (H) After the SR was loaded by a 5 x $10^{-7}$ M $Ca^{2+}$ solution $10^{-5}$ M GTP-$\gamma$-S was applied in the 2 mM EGTA 0 M $Ca^{2+}$ solution. $10^{-5}$ M GTP-$\gamma$-S induced a large transient contraction. (I) After the SR was depleted of $Ca^{2+}$ by 25 mM caffeine (not illustrated) $10^{-5}$ M GTP-$\gamma$-S and 1.8 x $10^{-7}$ M $Ca^{2+}$ were added. This procedure caused a tonic contraction which was inhibited reversibly by 2 x $10^{-5}$ M H-7. All experiments were done at 25°C using rabbit mesenteric artery. The experimental CSS solution contained 2 mM EGTA, 130 mM K propionate, 4 mM $MgCl_2$, 4 mM $Na_2ATP$, 10 mM CP, 0.1 mg/ml creatine phosphokinase, and 20 mM Tris maleate, pH = 6.8.

The α-adrenergic receptor mediated activation of a G protein could be bypassed through the use of guanosine-5'-O-(3-thiotriphosphate) (GTP-γ-S). $10^{-5}$ M GTP-γ-S induced large contractions, the pattern of which depended on the experimental conditions. The contraction was phasic followed by maintained tension when GTP-γ-S was applied to low $Ca^{2+}$ solution after the sarcoplasmic reticulum (SR) was loaded (Fig. 1F); it was transient when GTP-γ-S was added to a solution containing 2 mM EGTA and 0 mM $Ca^{2+}$ (Fig. 1H), and it was monophasic tonic when GTP-γ-S was added to a low $Ca^{2+}$ solution after the SR was depleted of $Ca^{2+}$ by caffeine (Fig. 1I). This latter contraction could be reversibly inhibited by 20 μM H-7, a concentration which had little effect on the force induced by $Ca^{2+}$ alone. In keeping with the known inability of GTPase to hydrolyze GTP-γ-S, its sensitizing effect was preserved after washout of the compound (Fig. 1G). These results suggest that receptor activation through the intermediate activation of a G protein and PKC (see also Fig. 3) increase the myofilament force sensitivity to cytoplasmic $Ca^{2+}$.

Figure 2 shows that addition of cAMP (0.3-30 μM) to α-toxin-permeabilized rat mesenteric artery dose-dependently inhibited the contraction induced by 0.5 μM $Ca^{2+}$ CSS solution (Fig. 2A). The possible contribution of $Ca^{2+}$ uptake by SR to cAMP-induced inhibition of contraction can be excluded, since 2 μM ionomycin, which depletes the intracellularly stored $Ca^{2+}$ (18), was present throughout this protocol. In the preliminary experiments, the treatment by 2 μM ionomycin completely abolished the caffeine induced contraction (data not shown). Thus, cAMP dose-dependently inhibited the contraction of α-toxin-permeabilized rat mesenteric artery at a fixed $Ca^{2+}$ concentration, indicating that cAMP decreases the $Ca^{2+}$ sensitivity of the contractile elements. As shown in Fig. 1B, forskolin, an activator of adenylate cyclase, reduced the contraction induced by 0.5 μM $Ca^{2+}$ CSS solution in the presence of 2 μM ionomycin, indicating that this compound also decreases the $Ca^{2+}$ sensitivity of the contractile elements. This observation indicates that adenylate cyclase remains intact inside the permeabilized cells and that the endogenously produced cAMP is sufficient to decrease the $Ca^{2+}$ sensitivity.

Application of cGMP (0.3-100 μM) to the permeabilized tissue dose-dependently inhibited the contraction induced by 0.5 μM $Ca^{2+}$ CSS solution (Fig. 2C), indicating that cGMP also decreases the $Ca^{2+}$ sensitivity of the contractile elements. 2 μM ionomycin was again present to exclude the contribution of SR $Ca^{2+}$ uptake. Figure 2D shows that 10 μM sodium nitroprusside, which is known to increase cGMP (19), could decrease the $Ca^{2+}$ sensitivity in the presence of 10 μM GTP. In the absence of GTP, only a small transient relaxation was observed by the application of sodium nitroprusside, indicating that GTP is required for this action probably to serve as the substrate of guanylate cyclase. These results show that guanylate cyclase remains intact in the permeabilized cells and that the concentration of cGMP produced endogenously by pharmacologic stimulation is sufficient to decrease the $Ca^{2+}$ sensitivity of contractile elements.

Figure 3A shows that 0.3 μM phorbol 12,13 dibutyrate (PDBu) induced a nearly maximal contraction in a 0.18 μM $Ca^{2+}$ CSS solution, which by itself did not increase tension at all. Since PDBu is known to activate PKC (4), it follows that PKC-mediated phosphorylation of as yet unidentified target sites increases $Ca^{2+}$ sensitivity of the contractile elements. The application of 30 μM cAMP after the PDBu-induced contraction reached a plateau value partially inhibited the muscle tension (Fig. 3A). Pre-

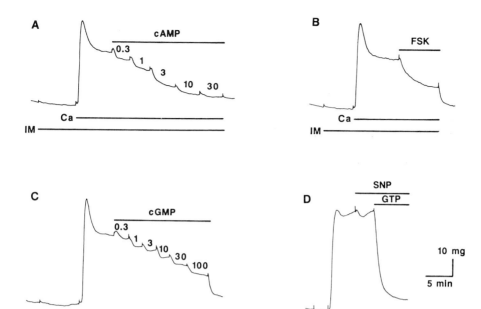

Figure 2: Effect of cAMP (A), forskolin (B), cGMP (C) and sodium nitroprusside (D) on $Ca^{2+}$-induced contraction of α-toxin-permeabilized rat mesenteric artery at 37°C. (A) cAMP dose-dependently inhibited the contraction induced by 0.5 μM $Ca^{2+}$ CSS solution (Ca). The concentrations (μM) of cAMP (cAMP) are illustrated above the trace. (B) 10 μM forskolin (FSK) inhibited the contraction induced by 0.5 μM $Ca^{2+}$ CSS solution (Ca). (C) cGMP dose-dependently inhibited the contraction induced by 0.5 μM $Ca^{2+}$ CSS solution (Ca). The concentrations (μM) of cGMP (cGMP) are illustrated above the trace. (D) 10 μM sodium nitroprusside (SNP) had little effect, but 10 μM sodium nitroprusside together with 10 μM GTP (GTP) inhibited the contraction induced by 0.5 μM $Ca^{2+}$ CSS solution (Ca). In all cases, 2 μM ionomycin (IM) was added to deplete SR (sarcoplasmic reticulum) $Ca^{2+}$. The composition of the CSS solution was the same as in Figure 1.

treatment with 30 μM cAMP inhibited the PDBu-induced contraction more effectively (Fig. 3B). The pretreatment with 30 μM cGMP also reduced the PDBu-induced contraction (Fig. 3C). These data indicate that the elevated $Ca^{2+}$ sensitivity of contractile elements induced by PKC can be partially reversed by PKA and PKG.

Figure 4 illustrates the effects of PDBu, cAMP and cGMP on the $pCa^{2+}$-tension relationship. PDBu caused a profound increase in $Ca^{2+}$ sensitivity of contractile elements, decreasing the half maximal effective concentration from $630 \pm 140$ nM (mean $\pm$ SD, n = 9) to $150 \pm 40$ nM (mean $\pm$ SD, n = 5, P < 0.01). On the other hand cAMP and cGMP decreased the $Ca^{2+}$ sensitivity with a reduction of the maximum response.

Besides agonists and second messengers, the myofilament sensitivity may also be manipulated by mere changes in substrate concentrations. Figure 5A shows the contraction induced by 145 mM $K^+$ PSS solution plus 10 μM NE in the intact tissue. The amplitude of this contraction was nearly the same as that of the 10 μM $Ca^{2+}$-induced contraction after α-toxin permeabilization (Fig. 5B), indicating that the majority of cells are permeabilized by α-toxin. Figure 5B-E shows the change of $Ca^{2+}$ sensitivity due to changing substrate conditions. Figure 5B shows the pCa$^{2+}$-tension relationship in the control condition (2 mM ATP, 20 mM CP, and 1.5 mM Mg$^{2+}$). If the ATP concentration was raised to 8 mM, while CP and Mg$^{2+}$ concentrations were unchanged (20 mM and 1.5 mM, respectively), the pCa$^{2+}$-tension relationship was shifted to the left (Fig. 5D). If the increase of Ca$^{2+}$ sensitivity observed in Figure 5D is due to the rise in ATP concentration, one would expect that the Ca$^{2+}$ sensitivity should be decreased by lowering the CP concentration, since less ATP will be regenerated from ADP and CP catalyzed by endogenous creatine phosphokinase. However, as shown in Fig. 5E, reduction of the CP concentration from the condition described in Fig. 5D (20 mM) to 10 mM,

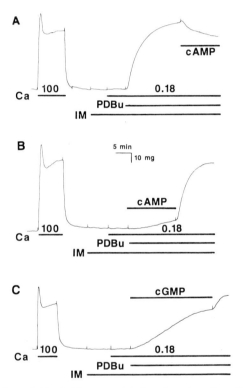

Figure 3: PDBu (phorbol 12,13 dibutyrate)-induced contraction (A) and the effect of cAMP (A, B) and cGMP (C) on PDBu-induced contraction at a fixed Ca$^{2+}$ concentration of 0.18 µM in α-toxin-permeabilized rat mesenteric artery at 37°C. (A) After the contraction induced by 100 µM Ca$^{2+}$ CSS solution (illustrated as Ca and 100), 0.18 µM Ca$^{2+}$ CSS solution (illustrated as 0.18) was added. This Ca$^{2+}$ solution did not contract the tissue at all. The application of 0.3 µM PDBu in addition to 0.18 µM Ca$^{2+}$ CSS solution induced a large contraction. After this contraction reached plateau, 30 µM cAMP (cAMP) was added. This procedure partially relaxed the tissue. (B) Pretreatment by 30 µM cAMP (cAMP) inhibited the 0.3 µM PDBu-induced contraction. This inhibition could be washed out. (C) Pretreatment by 30 µM cGMP (cGMP) also inhibited the 0.3 µM PDBu-induced contraction. This inhibition could be also washed out. In A, B and C, 2 µM ionomycin (IM) was added to exclude the contribution of SR Ca$^{2+}$ uptake. Results illustrated are representative of three independent experiments. A, B and C were obtained in different tissues from the same rat. The composition of the CSS solution was the same as in Figure 1.

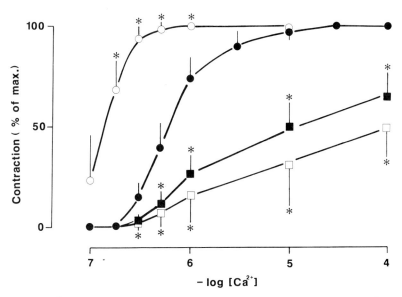

Figure 4: $Ca^{2+}$-force relationship in the absence ($\bullet$: n = 9) or presence of 30 $\mu$M cAMP ($\square$: n = 8), 30 $\mu$M cGMP ($\blacksquare$: n = 5) and 0.3 $\mu$M PDBu ($\bigcirc$: n = 5), in $\alpha$-toxin-permeabilized rat mesenteric artery at 37°C. Values are expressed as percentage of maximum contraction. In case of cAMP and cGMP, contractions induced by 100 $\mu$M $Ca^{2+}$ CSS solution without these agents were defined as 100%. Experiments were carried out at 37°C. Vertical bars represent standard deviation. The composition of the CSS solution was the same as in Figure 1. * significantly different from the values in control, P< 0.01.

while other components were kept the same, caused a further leftward shift of the pCa2+-tension curve. From these results, we considered that the leftward shift of the pCa2+-tension curve might be due to the accumulation of ADP. In other words, CP functions not only to regenerate ATP but also to reduce the ADP concentration. This hypothesis was confirmed in Fig. 5C, where 100 $\mu$M adenosine-5'-O-(2-thiodiphosphate) (ADP-$\beta$-S), a non-hydrolyzable ADP analog, caused a similar leftward shift of the pCa2+-tension curve, while the concentrations of ATP and CP were unchanged (2 mM and 20 mM, respectively).

It was also apparent that relaxation of the fiber becomes much slower or even incomplete at higher ATP and/or lower CP concentrations, conditions under which ADP may more easily accumulate. As shown in Fig. 5F, 0.1 $\mu$M Ca2+ CSS solution which contained 2 mM ATP and 20 mM CP did not cause contraction and rapidly relaxed the fiber to the basal level, even if it was preincubated by 10 $\mu$M Ca2+ CSS solution (first contraction). However, under the conditions of 4 mM ATP and 10 mM CP (second contraction in Fig. 5F) or 8 mM ATP and 10 mM CP (third contraction), 0.1 $\mu$M Ca2+ solution only slowly relaxed the fiber precontracted by 10 $\mu$M Ca2+ solution. The steady state force induced by 0.1 $\mu$M Ca2+ solution after the 10 $\mu$M Ca2+ induced contraction appeared to be higher than that before the 10 $\mu$M Ca2+-induced contraction. This slow relaxation and hysteresis in the latter condition may be also due to the accumulation of ADP. The increased concentration of ADP could lead to accumulation of actomyosin-ADP complex as supported by the results of Fig. 6A, showing that a large contraction can be induced by a 0.18 $\mu$M Ca2+ solution containing 8 mM

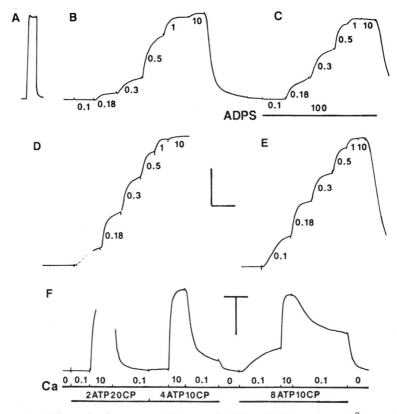

Figure 5: Effect of substrate condition and ADP-β-S (C) on the pCa$^{2+}$-tension relationship (B-E) and relaxation rate (F) in α-toxin-permeabilized rabbit mesenteric artery at 37°C. (A) Contraction of the intact tissue induced by 145 mM K$^+$ PSS solution containing 10 μM NE. (B) pCa$^{2+}$-tension relationship in the control condition, 2 mM ATP, 20 mM CP, and 1.5 mM Mg$^{2+}$. (C) Effect of 100 μM ADPβS (ADPS) on pCa$^{2+}$-tension relationship in the control condition. (D) pCa$^{2+}$-tension relationship under the condition of 8 mM ATP, 20 mM CP, and 1.5 mM Mg$^{2+}$. (E) pCa$^{2+}$-tension relationship under the condition of 8 mM ATP, 10 mM CP, and 1.5 mM Mg$^{2+}$. (F) First, 0.1 μM Ca$^{2+}$ solution was applied and then 10 μM Ca$^{2+}$ solution was applied. After this contraction reached steady state, the tissue was then relaxed by re-adding 0.1 μM Ca$^{2+}$ solution. This protocol was performed in the 2 mM ATP, 20 mM CP condition (first contraction, illustrated "2ATP20CP"), the 4 mM ATP, 10 mM CP condition (second contraction, "4ATP10CP") and in 8 mM ATP, 10 mM CP condition (third contraction, "8ATP10CP"). Ca$^{2+}$ concentrations are given under the trace in μM. The control CSS solution contained 2 mM EGTA, 130 mM K propionate, 3.1 mM MgCl$_2$, 2 mM Na$_2$ATP, 20 mM CP, and 20 mM Tris maleate, pH = 6.8. In case of changing the ATP concentration, the MgCl$_2$ concentration was also changed as to keep the Mg$^{2+}$ concentration constant, using the binding constant 10$^4$ for MgATP complex. Vertical and horizontal calibrations represent 50 mg and 5 min, respectively. Traces shown are representative of 3 experiments.

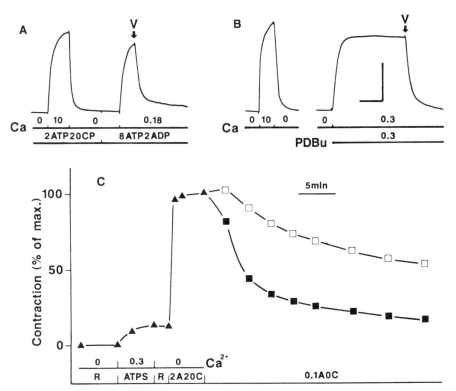

Figure 6: Effect of 1 mM vanadate (V) on a 2 mM Mg ADP (A) and a 0.3 μM PDBu
(B) induced contraction and low substrate (0.1 mM ATP) solution-induced
relaxation (C) in α-toxin-permeabilized rabbit mesenteric artery at 37°C. (A)
First, a control 10 μM $Ca^{2+}$-induced contraction was recorded in 2 mM ATP, 20 mM
CP condition. Then, 0.18 μM $Ca^{2+}$ solution containing 8 mM ATP and 2 mM
MgADP was added. After this contraction reached steady state, 1 mM vanadate (V)
was applied. (B) First, a control 10 μM $Ca^{2+}$-induced contraction was recorded.
Then, a 0.3 μM $Ca^{2+}$ solution containing 0.3 μM PDBu was added. After this
contraction reached steady state, 1 mM vanadate (V) was applied. The substrate
condition was the same as control (2 ATP, 20 CP) throughout the record. (C) First,
the tissues were exposed to rigor solution which contains no $Ca^{2+}$, no ATP and no
CP (illustrated as "R"). After this procedure, the tissues were exposed to 0.3 μM
$Ca^{2+}$ solution containing 0.1 mM ATP-γ-S for 5 min (illustrated as "ATPS").
Then $Ca^{2+}$ and ATP-γ-S were washed out by rigor solution and $Ca^{2+}$ free solution
containing 2 mM ATP and 20 mM CP was added. After the contractions induced by
these procedures reached steady state, low substrate solutions (0.1 mM ATP)
containing 0 M $Ca^{2+}$ (□; n = 3) and 10 μM $Ca^{2+}$ solution (■; n = 4) were added. SD
bars were omitted for clarity, but averaged 7% for open squares and 6% for closed
squares. In order to remove CP completely, washout by the same solution were
carried out 3 times during relaxation. $Ca^{2+}$ concentrations in μM and substrate
conditions are illustrated under the trace. On the vertical axis, contractions were
plotted as percentages of maximum. The horizontal calibration represent 5 min.
The composition of the control CSS solution was the same as in Figure 5.

MgATP and 2 mM MgADP and that this contraction was almost completely inhibited by vanadate (see discussion). Vanadate had the same relaxing effect on a contraction activated by PDBu in the presence of a low myoplasmic $Ca^{2+}$ concentration (Fig. 6B). The lower curves of Fig. 6 present evidence for a second $Ca^{2+}$ regulatory mechanism present in smooth muscle which does not involve MLCK. First, the tissue was exposed to rigor solution which contains no ATP, no CP and no $Ca^{2+}$ for 5 min. This exposure to rigor solution did not induce any contraction. After this treatment a 0.3 μM $Ca^{2+}$ solution containing 0.1 mM ATP-γ-S was added for 5 min. This procedure caused a small contraction. Then, the tissue was washed again with rigor solution. After all these procedures, $Ca^{2+}$ free solution containing 2 mM ATP and 20 mM CP was added. This treatment induced a maximum contraction as indicated by the observation that a subsequent addition of 10 μM $Ca^{2+}$ solution did not cause additional tension development (data not shown). This contraction was attributed to the irreversible thiophosphorylation of MLC, which is known to be resistent to phosphatase (20). Thus, the phosphorylation level of MLC should be fixed at a high level after the exposure to ATP-γ-S. The subsequent application of a $Ca^{2+}$ free solution containing 0.1 mM ATP caused a very slow relaxation, which could be dramatically accelerated by the addition of 10 μM $Ca^{2+}$.

DISCUSSION

Although α-toxin was used more than 10 years ago to permeabilize smooth muscle (20) it was not recognized until 1988 that this procedure allowed the study of receptor mediated events at fixed levels of $[Ca^{2+}]_i$ (16,21). This proteinaceous toxin (MW 33,000) is produced by Staphylococcus aureus and forms pores of 2-3 nm diameter when it aggregates into hexamers in the cell membrane (21). The limited pore size allows equilibration of the cytoplasm with inorganic ions and small molecules contained in the experimental solutions, but prevents permeation by proteins. Thus vital components such as calmodulin and enzymes are preserved in the cytoplasm, while α-toxin itself is prevented from entering the cells and permeabilzing the intracellular organelles. This last point is illustrated by the large transient $Ca^{2+}$ release contraction evoked by GTP-γ-S (Fig. 1H). Permeabilization to $Ca^{2+}$ and adenine and guanine nucleotides is illustrated by the contractile responses to submicromolar concentrations of $Ca^{2+}$ which are dependent on the presence of ATP and CP, the requirement for GTP in the NE induced contractions and the responsiveness to GTP-γ-S. In this preparation it was possible to clamp the $[Ca^{2+}]_i$ and observe that NE or phenylephrine through the intermediate activation of a G protein was able to increase the smooth muscle myofilament sensitivity to $Ca^{2+}$ (16,22). The data further indicate that second messengers (probably through subsequent activation of specific protein kinases) directly regulate smooth muscle contractile elements. Support for this hypothesis had been mustered by the recent advances in simultaneous measurements of tension and $[Ca^{2+}]_i$, using the luminescent protein aequorin and the fluorescent $Ca^{2+}$ binding dyes, quin 2 and fura 2. Morgan and Morgan (2) first reported that α-adrenergic activation of ferret portal vein caused a transient peak aequorin luminescence which returned close to baseline while tension was increased in a tonic manner. In addition the same authors noted that isoprenaline, papaverine and forskolin, which are known to increase cellular cAMP levels, produced either an increase or no change in $[Ca^{2+}]_i$ while the arterial smooth muscle relaxed (23). Takuwa et al.(24) observed that isoproterenol, forskolin and vasoactive intestinal peptide, which are known to stimulate adenylate cyclase, increased $[Ca^{2+}]_i$ without any tension development in bovine tracheal smooth muscle. Takuwa and Rasmussen (25) reported that atrial natriuretic peptide, which is known to increase cellular cGMP level (26), did not inhibit the $Ca^{2+}$ transient induced by histamine or angiotensin II, while this peptide inhibited the contractile response of the rabbit aorta. Felbel $et\ al$ (27) also reported that isoproterenol increased $[Ca^{2+}]_i$ and that pretreatment by isoproterenol, forskolin, or 8 Br-cAMP did not inhibit the carbachol-induced increase in $[Ca^{2+}]_i$ of bovine tracheal smooth muscle. However, nitrovasodilators such as nitro-glycerine (28) and sodium nitroprusside (23,29), which are known to increase cellular cGMP level (19) and 8 Br-cGMP (30) are reported to decrease $[Ca^{2+}]_i$, although it was stressed that sodium nitroprusside was more effective in relaxing vascular smooth muscle than in decreasing $[Ca^{2+}]_i$ (23,29). Furthermore, Jiang and Morgan (31) reported that phorbol ester induced smooth muscle contraction with no change in $[Ca^{2+}]_i$. Although these observations supported direct control by second messengers of smooth muscle contractile elements, they were not accepted as proof (14) probably due to technical reservations. Compartmentalization of the intracellular $Ca^{2+}$ indicators has not been completely ruled out while the possible presence of intracellular $Ca^{2+}$ gradients could complicate measurement of $[Ca^{2+}]_i$ near the myofilaments (32).

A more direct biochemical approach was taken earlier by Adelstein and collaborators, who showed that PKA mediated phosphorylation of MLCK lowered its affinity for the $Ca^{2+}$ calmodulin complex (12). The relaxing effect related to this mechanism was confirmed by Kerrick and Hoar (33) who found that addition of the catalytic subunit of PKA inhibits $Ca^{2+}$ activated force of skinned chicken gizzard smooth muscle. This effect is inhibited by calmodulin which is often added to skinned smooth muscle preparations in order to enhance sensitivity to $Ca^{2+}$ (15,34,35). The introduction of this variable unfortunately obscures the physiological relevance of Adelstein's proposed mechanism for cAMP induced smooth muscle relaxation (13,14). It has indeed been argued that under physiological conditions the cells contain sufficient calmodulin to minimize the effect of MLCK phosphorylation (13). In $\alpha$-toxin-permeabilized preparations, calmodulin may be maintained at physiological level, since the addition of exogenous calmodulin (5 $\mu$M) had no effect on $Ca^{2+}$-induced contraction, indicating that calmodulin was not permeable through the pores. Thus the observation that cAMP and cGMP induced relaxation of the $\alpha$-toxin treated smooth muscle at constant $[Ca^{2+}]$ supports a physiological function for down regulation of myofilament $Ca^{2+}$ sensitivity. Furthermore, the observation that endogenously produced cAMP or cGMP were also effective in lowering myofilament $Ca^{2+}$ sensitivity supports the hypothesis that this mechanism of relaxation functions both under physiologic and therapeutic conditions. Since cAMP and cGMP by themselves did not change the $Ca^{2+}$ sensitivity of contractile elements in saponin-treated skinned fiber (15,36), probably due to the loss of protein kinases from the tissue and since the addition of PKA or PKG decreased the $Ca^{2+}$ sensitivity (15,33,35), it is likely that these second messenger effects are mediated through activation of specific protein kinases. This does however not imply that the above is the sole mechanism for cyclic nucleotide induced relaxation since in intact cells their ability to lower $[Ca^{2+}]_i$ has been well documented (13,14). The above leaves little doubt that physiological mediators can both up- and down-regulate the $Ca^{2+}$ sensitivity of vascular smooth muscle. However, the mechanisms involved require more detailed investigation.

Our finding that direct stimulation of PKC by PDBu initiated a large and rapid contraction at 37°C is perhaps the strongest evidence that this enzyme is involved in the observed up regulation of myofilament $Ca^{2+}$ sensitivity. Other supporting evidence includes the selective inhibition of the GTP-γ-S contraction by H-7 and the slow phorbol ester induced contractions of intact arteries, which are not accompanied by increases in $[Ca^{2+}]_i$ (31,37). In addition, we have also presented evidence for a role of relatively low concentrations of ADP in elevating myofilament $Ca^{2+}$ sensitivity (17).

Studies using isolated contractile proteins have shown that actomyosin hydrolyzes ATP via a series of reactions (38,39,40), a simplified version of which is illustrated in scheme 1. Since AM-ADP is the force generating intermediate in the normal crossbridge cycle, this scheme predicts that increasing the concentration of ADP should increase the population of the AM-ADP complex and the tension generated at a fixed intermediate $Ca^{2+}$ concentration.

This prediction has been verified by Ventura-Clapier et al in the heart (41), Hoar et al in skeletal muscle (42) and Kerrick and Hoar in smooth muscle (43). The latter authors pointed out that the ADP effect in smooth muscle would be independent of MLC phosphorylation and would slow ATP consumption and shortening velocity, properties characteristic of the "latch" state. However, millimolar concentrations of ADP were required to generate force in chicken gizzard smooth muscle permeabilized by Triton X. Thus, the reservations in explaining smooth muscle "latch" contracture on the basis of ADP accumulation were that unphysiologically high ADP concentrations were required and that this effect was not specific for smooth muscle. However, Fig. 5 shows that addition of 100 μM ADP-β-S or increasing the ADP concentration by merely increasing the ATP/CP ratio dramatically increases the $Ca^{2+}$ sensitivity of α-toxin permeabilized smooth muscle (17). It can thus be concluded that this preparation is very sensitive to ADP bringing the effective concentration (100 μM) within the range of ADP concentrations encountered under physiological conditions. Krisanda and Paul (44) reported a value of 0.27 μmol ADP/g wet weight or approximately 400 μmol/liter cytoplasm (assuming 32% extracellular space (45)) in high $K^+$ contracted smooth muscle, which was significantly higher than in resting muscle. Kushmerick et al (46) chemically measured an ADP concentration of 340 μM in perfused rabbit bladder, while NMR analysis of smooth muscle [ADP] yielded a value of 0.16 μmol/g wet weight or 235 μM (47). Thus, it seems plausible to explain the hysteresis observed during $K^+$ stimulation and subsequent relaxation of the intact ferret aorta by DeFeo and Morgan (48) by an accumulation of ADP in the intact working muscle. However, ADP accumulation probably does not account for the difference in $[Ca^{2+}]_i$ sensitivity measured in the ferret portal vein between phenylephrine stimulation and high $K^+$ depolarization (48). Neither could it account for the PDBu initiated increase in $Ca^{2+}$ sensitivity of the α-toxin permeabilized rabbit mesenteric artery under conditions which prevent ADP accumulation (Fig. 4)

If a latch-like state can be induced by reaction "1" in Scheme 1, one would also predict that the inhibition of ADP release from the AM-ADP complex (inhibition of reaction "2" in the Scheme 1) would increase the population of AM-ADP, slow down the crossbridge cycling rate, and induce a latch-like state, regardless of the energy state of the smooth muscle. This idea has been proposed previously by Lash et al (49) in a study of the function of caldesmon in a cell free in-vitro system. Recent biochemical

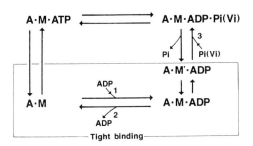

Scheme 1: Simplified scheme of actomyosin ATPase (ref. 38-40). Abbreviations used are as follows: A: actin, M: myosin, Pi: phosphate, Vi: vanadate.

studies have shown that, under low $Ca^{2+}$ condition, caldesmon inhibits actomyosin ATPase activity (50), that this inhibition can be reversed by the addition of $Ca^{2+}$ and calmodulin, and that the ATPase inhibition is associated with an increase in the affinity of smooth muscle heavy meromyosin (HMM) for actin filament (49,51,52).

Two different hypotheses have been put forward to explain how caldesmon inhibits actomyosin ATPase and causes tight binding between actin and myosin. The first holds that caldesmon inhibits release of ADP from the acto-HMM-ADP complex causing accumulation of tightly bound crossbridges and also inhibits actomyosin ATPase activity since ADP release is the rate limiting step (49). In support of this hypothesis, Marston (53) reported that caldesmon inhibits actomyosin ATPase by inhibiting the product release from the AM-ADP-Pi complex. The second hypothesis states that caldesmon cross-links actin and myosin filaments and by so doing inhibits crossbridge cycling (51,52).

Although it is clear that the exa   role played by caldesmon in latch formation remains unresolved (54), the data gathered on the α-toxin permeabilized smooth muscle favor a pivotal role for the accumulation of AM-ADP. This preparation has a very high affinity for ADP and the accumulation of AM-ADP by increments of ADP increases force as predicted by *in vitro* studies (49). The hypothesis that accumulation of AM-ADP is characteristic of the latch like state reported in this presentation is further supported by the relaxant effects of vanadate on contractions induced by ADP or by phorbol ester. Vanadate is an analog of phosphate and is reported to cause relaxation by complexing with AM-ADP (reaction "3" in Scheme 1) (55). However, since the contractile effect of PDBu was observed with 20 mM creatine phosphate and 2 mM ATP, it is more likely to be due to an inhibition of ADP release than to a mass action effect of accumulated ADP.

Our data could be explained according to the above hypothesis (see also ref. 17), if PKC mediated phosphorylation of caldesmon (56,57) decreased its sensitivity to $Ca^{2+}$, allowing inhibition of ADP release from the AM-ADP complex. The final figure (6,C) indicates that a second $Ca^{2+}$ regulated system possibly caldesmon may be involved in the maintenance of smooth muscle tension, since a high $[Ca^{2+}]_i$ was capable of inducing relaxation in spite of maintained thiophosphorylation of the myosin light chains. In conclusion, our data are compatible with the hypothesis that latch type contraction is due to the accumulation of AM-ADP complexes due to either accumulation of ADP or caldesmon mediated inhibition of ADP release from the tightly bound crossbridges and that these processes may be modulated by second messengers.

ACKNOWLEDGMENTS

We are particularly indebted to Dr. R.J. Hohman for supplying us with α-toxin. This work was supported by NIH grant HL 35657 and a postdoctoral fellowship to J.N. from the American Heart Association, Florida Affiliate. The authors also thank Gerry Trebilcock for her secretarial assistance.

REFERENCES

1. Somlyo AP. Excitation-contraction coupling and the ultrastructure of smooth muscle. *Circ Res* 57: 497, 1985.
2. Morgan JP, Morgan KG. Vascular smooth muscle: The first recorded $Ca^{2+}$ transients. *Pflugers Arch* 395: 75, 1982.
3. Morgan JP, Morgan KG. Stimulus-specific patterns of intracellular calcium levels in smooth muscle of the ferret portal vein. *J Physiol (Lond)* 351: 155, 1984.
4. Nishizuka Y. Studies and perspectives of protein kinase C. *Science* 233: 305, 1986.
5. Baraban JM, Gould RJ, Peroutka SJ, Snyder SH. Phorbol ester effects on neurotransmission: Interaction with neurotransmitters and calcium in smooth muscle. *Proc Natl Acad Sci USA*, 82: 604, 1985.
6. Danthuluri NR, Deth RC. Phorbol ester-induced contraction of arterial smooth muscle and inhibition of α-adrenergic response. *Biochem Biophys Res Comm* 125: 1103, 1984.
7. Forder J, Scriabine A, Rasmussen H. Plasma membrane calcium flux, protein kinase C activation and smooth muscle contraction. *J Pharmacol Exp Ther* 235: 267, 1985.
8. Rasmussen M, Forder J, Kojima I, Scriabine A. TPA-induced contraction of isolated rabbit vascular smooth muscle. *Biochem Biophys Res Comm* 122: 776, 1984.
9. Khalil RA, van Breemen C. Sustained contraction of vascular smooth muscle: Calcium influx or C-kinase activation? *J Pharmacol Exp Ther* 244: 537, 1988.
10. Itoh T, Kubota Y, Kuriyama H. Effects of phorbol ester on acetylcholine-induced $Ca^{2+}$ mobilization and contraction in the porcine coronary artery. *J Physiol (Lond)* 397: 401-419, 1988.
11. Nishimura J, van Breemen C. Direct regulation of smooth muscle contractile elements by second messengers. *Biochem Biophys Res Comm* 163: 929, 1989.
12. Adelstein RS, Conti MA, Hathaway DR, Klee CB. Phosphorylation of smooth muscle myosin light chain kinase by the catalytic subunit of adenosine 3':5'-monophosphate-dependent protein kinase. *J Biol Chem* 253: 8347, 1978.
13. Kamm KE, Stull JT. The function of myosin and myosin light chain kinase phosphorylation in smooth muscle. *Annu Rev Pharmacol Toxicol* 25: 593, 1985.
14. Kamm KE, Stull JT. Regulation of smooth muscle contractile elements by second messengers. *Annu Rev Physiol* 51: 299:313, 1989.
15. Itoh T, Kanmura Y, Kuriyama H, Sasaguri T. Nitroglycerine- and isoprenaline-induced vasodilation: assessment from the actions of cyclic nucleotides. *Br J Pharmacol* 84: 393, 1985.
16. Nishimura J, Kolber M, van Breemen C. Norepinephrine and GTP-γ-S increase myofilament $Ca^{2+}$ sensitivity in α-toxin permeabilized arterial smooth muscle. *Biochem Biophys Res Comm* 157: 677, 1988.

17. Nishimura J, van Breemen C. Possible involvement of actomyosin ADP complex in regulation of $Ca^{2+}$ sensitivity in α-toxin permeabilized smooth muscle. *Biochem Biophys Res Comm* 165: 408, 1989.
18. Fujiwara T, Itoh T, Kubota Y, Kuriyama H. Effect of guanosine nucleotides on skinned smooth muscle tissue of the rabbit mesenteric artery. *J Physiol (Lond)* 408: 535, 1989.
19. Schultz KD, Schultz K, Schultz J. Sodium nitroprusside and other smooth muscle-relaxants increase cyclic GMP levels in rat ductus deferens. *Nature* 265: 750, 1977.
20. Cassidy P, Hoar PE, Kerrick WGL. Irreversible thiophosphorylation and activation of tension in functionally skipped rabbit ileum strips by [$^{35}$S]ATPγS. *J Biol Chem* 254: 11148, 1979.
21. Hohman RJ. Aggregation of IgE receptors induces degranulation in rat basophilic leukemia cells permeabilized with α-toxin from Staphylococcus aureus. *Proc Natl Acad Sci USA* 85: 1624, 1988.
22. Kitazawa T, Kobayashi S, Horiuchi K, Somlyo AV, Somlyo AP. Receptor coupled, permeabilized smooth muscle: role of the phosphatidylinositol cascade, G-proteins and modulation of the contractile response to $Ca^{2+}$. *J Biol Chem* 264: 5339, 1989.
23. Morgan JP, Morgan KG. Alteration of cytoplasmic ionized calcium level in smooth muscle by vasodilators in the ferret. *J Physiol (Lond)* 357: 539, 1984.
24. Takuwa Y, Takuwa N, Rasumussen H. The effects of isoproterenol on intracellular calcium concentration. *J Biol Chem* 263: 762, 1988.
25. Takuwa Y and Rasmussen H. Measurement of cytoplasmic free $Ca^{2+}$ concentration in rabbit aorta using the photoprotein, aequorin. *J Clin Invest* 80: 248, 1987.
26. Waldman SA, Rapoport RM, Murad F. Atrial natriuretic factor selectively activates particular guanylate cyclase and elevates cyclic GMP in rat tissues. *J Biol Chem* 259: 14332, 1984.
27. Felbel J, Trockur B, Ecker T, Landgraf W, Hofmann F. Regulation of cytosolic calcium by cAMP and cGMP in freshly isolated smooth muscle cells from bovine trachea. *J Biol Chem* 263: 16764, 1988.
28. Kobayashi S, Kanaide H, Nakamura M. Cytosolic free calcium transient in cultured smooth muscle cells: Microfluorometric measurements. *Science* 229: 553, 1985.
29. Karaki H, Sato K, Ozaki H, Murakami K. Effects of sodium nitroprusside on cytosolic calcium level in vascular smooth muscle. *Eur J Pharmacol* 156: 259, 1988.
30. Kai H, Kanaide H, Matsumoto T, Nakamura M. 8-Bromoguanosine 3':5'-cyclic monophosphate decreases intracellular free calcium concentrations in cultured vascular smooth muscle cells from rat aorta, *FEBS Lett* 221: 284, 1987.
31. Jiang MJ, Morgan KG. Intracellular calcium levels in phorbol ester-induced contraction of vascular muscle. *Am J Physiol* 253: H1365, 1987.
32. Rembold CM, Murphy RA. Myoplasmic $Ca^{2+}$ determines myosin phosphorylation in agonist-stimulated swine arterial smooth muscle. *Circ Res* 63: 593, 1988.
33. Kerrick WGL, Hoar PE. Inhibition of smooth muscle tension by cyclic AMP-dependent protein kinase. *Nature (Lond)* 292: 253, 1981.
34. Cassidy PS, Kerrick WGL, Hoar PE, Malencik DA. Exogenous calmodulin increases $Ca^{2+}$ sensitivity of isometric tension activation and myosin phosphorylation in skinned smooth muscle. *Pflügers Arch* 392: 115, 1981.

35. Ruegg JC, Paul RJ. Vascular smooth muscle calmodulin and cyclic AMP-dependent protein kinase alter calcium sensitivity in porcine carotid skinned fibers. *Circ Res* 50: 394, 1982.

36. Saida K, van Breemen C. Cyclic AMP modulation of adrenoceptor mediated arterial smooth muscle contraction. *J Gen Physiol* 84: 307, 1984.

37. Nishimura J, Khalil RA, van Breemen C. Evidence for increased myofilament $Ca^{2+}$ sensitivity in norepinephrine-activated vascular smooth muscle. *Am J Physiol,* in press.

38. Eisenberg E, Greene LE. The relation of muscle biochemistry to muscle physiology. *Annu Rev Physiol* 42: 293, 1980.

39. Sleep JA, Hutton RL. Exchange between inorganic phosphate and adenosine 5'-triphosphate in the medium by actomyosin subfragment 1. *Biochemistry* 19: 1276, 1980.

40. Sellers JR. mechanism of the phosphorylation-dependent regulation of smooth muscle heavy meromyosin. *J Biol Chem* 260: 15815, 1985.

41. Ventura-Clapier R, Mekhfi H, Vassort G. Role of creatine kinase in force development in chemically skinned rat cardiac muscle. *J Gen Physiol* 89: 815, 1987.

42. Hoar PE, Mahoney CW, Kerrick WGL. $MgADP^-$increases maximum tension and $Ca^{2+}$ sensitivity in skinned rabbit soleus fibers. *Pflugers Arch* 410: 30-36, 1987.

43. Kerrick WGL, Hoar PE. Non-$Ca^{2+}$-activated contraction in smooth muscle, in: *Regulation and Contraction of Smooth Muscle*, M.J. Siegmanm A.P. Somlyo, and N.L. Stephens, eds., A.R. Liss, New York, 1987.

44. Krisanda JM, Paul RJ. Phosphagen and metabolite content during contraction in porcine carotid artery. *Am J Physiol* 244: C385, 1983.

45. Gabella G. The force generated by a visceral smooth muscle. *J Physiol (Lond)* 263: 199, 1976.

46. Kushmerick MJ, Dillon PF, Meyer RA, Brown TR, Krisanda JM, Sweeney HE. $^{31}P$ NMR spectroscopy, chemical analysis, and free $Mg^{2+}$ of rabbit bladder and uterine smooth muscle. *J Biol Chem* 261: 14420, 1986.

47. Yoshizaki K, Radda GK, Inubushi T, Chance B. $^1H$- and $^{31}P$-NMR studies on smooth muscle of bullfrog stomach. *Biochim Biophys Acta* 928: 36, 1987.

48. DeFeo TT, Morgan KG. Calcium-force relationships as detected with aequorin in two different vascular smooth muscles of the ferret. *J Physiol (Lond)* 369: 269, 1985.

49. Lash JA, Sellers JR, Hathaway DR. The effects of caldesmon on smooth muscle heavy actomeromyosin ATPase activity and binding of heavy meromyosin to actin. *J Biol Chem* 261: 16155, 1986.

50. Sobue K, Muramoto Y, Fujita M, Kakiuchi S. Purification of a calmodulin-binding protein from chicken gizzard that interacts with F-actin. *Proc Natl Acad Sci USA* 78: 5652, 1981.

51. Ikebe M, Reardon S. Binding of caldesmon to smooth muscle myosin. *J Biol Chem* 263: 3055, 1988.

52. Sutherland C, Walsh MP. Phosphorylation of caldesmon prevents its interaction with smooth muscle myosin. *J Biol Chem* 264: 578, 1989.

53. Marston SB. Aorta caldesmon inhibits actin activation of thiophosphorylated heavy meromyosin $Mg^{2+}$-ATPase activity by slowing the rate of product release. *FEBS Lett* 238: 147, 1988.

54. Marston SB. What is the latch? New ideas about tonic contraction in smooth muscle. *J Musc Res Cell Motility* 10: 97, 1989.

55. Dantzig JA, Goldman YE. Suppression of muscle contraction by vanadate: Mechanical and ligand binding studies on glycerol-extracted rabbit fibers. *J Gen Physiol* 86: 305, 1985.
56. Umekawa H, Hidaka H. Phosphorylation of caldesmon by protein kinase C. *Biochem Biophys Res Comm* 132: 56, 1985.
57. Park S, Rasmussen H. Carbachol-induced protein phosphorylation changes in bovine tracheal smooth muscle. *J Biol Chem* 261: 15734, 1986.

# POTASSIUM CHANNEL ACTIVATORS IN
# VASCULAR SMOOTH MUSCLE

Robert H. Cox

Bockus Research Institute
The Graduate Hospital and
Department of Physiology
The University of Pennsylvania
Philadelphia, PA

## INTRODUCTION

Ion channels play an important role in the function of all cells. They are involved in the determination of resting membrane potential, intracellular ion levels, action potential generation, and stimulus-response coupling. Over the last ten years, interest in ion channels as targets for therapeutic drug development has steadily increased. The $Na^+$ channel was the first recognized as the site of action of therapeutically important compounds such as the local anesthetics. Over the last decade, $Ca^{2+}$ channels have been the prominent site for the development of new modulatory agents. This activity has witnessed the development of numerous compounds with varying tissue specificity that modulate $Ca^{2+}$ channel function. More recently, $K^+$ channels have also become an important site of interest for the development of new therapeutic compounds (1). While $K^+$ channel blockers have been available for many years, compounds that appear to function as $K^+$ channel activators have been synthesized only recently.

The resting membrane potential of most cells including striated and smooth muscles is dominated by $K^+$ channel permeability and the transmembrane $K^+$ distribution. Other ion channels ($Na^+$ and $Cl^-$) generally make a smaller contribution to resting membrane potential. $K^+$ channel activators by their ability to increase $K^+$ permeability of membranes have the capacity to shift resting membrane potential toward the $K^+$ equilibrium potential which usually represents a hyperpolarizing effect. This action in vascular smooth muscle is relatively more pronounced because the contributions of the permeabilities of $Na^+$ and $Cl^-$ to membrane potential compared to that of $K^+$ are larger than in most other cells (2). The effect of increasing $K^+$ channel permeability in vascular smooth muscle is to cause relaxation of existing contractions or inhibition of the subsequent action of vasoconstrictor agents. This is due to the close relation between membrane potential and force development (or tone) in vascular smooth muscle (3).

In recent years, a number of agents of substantially different chemical structure that have hypotensive actions *in vivo* have been suggested to act as $K^+$ channel activators. These include cromakalim, diazoxide, minoxadil, nicorandil and pinacidil. The following will review the functional properties of these compounds both *in vivo* and *in vitro*, and will discuss the evidence for their mechanism of action as well as questions which remained unanswered regarding these proposed mechanisms. The discussion will focus on cromakalim (BRL-34915) since a larger volume of research exists for this compound compared to others.

## BASIC PROPERTIES

### Hemodynamics

Cromakalim and other K+ channel activators have a profound effect on blood pressure when administered orally or intravenously (1,4-9). With intravenous administration in anesthetized animals, blood pressure decreases rapidly to a minimum about 20 minutes after administration in a dose dependent manner as shown in figure 1 (6).

With oral administration in conscious animals, blood pressure also decreases to a minimum value in about 20 minutes but recovers partially over a short time course and is well maintained for several hours (4,6). These effects are observed in both normotensive subjects as well as in subjects with genetic or experimental forms of hypertension. The effects of cromakalim on blood pressure are not altered by antagonists to muscarinic, histamine, $\alpha_2$- and $\beta_2$-adrenergic receptors, or mediated via cyclooxygenase pathways involved in the production of vasodilator prostaglandins (6). Cromakalim's hypotensive action is about 10-30 times more potent than that of nifedipine (4) and about three-fold more potent than pinacidil (1).

Cromakalim appears to preferentially dilate coronary, gastro-intestinal, and cerebral vasculatures as compared to those of the kidney and skeletal muscle where it is less effective (4,7). This profile of activity is

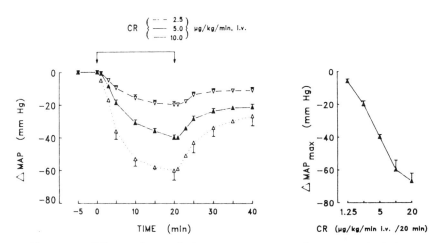

Figure 1: Effects of cromakalim on mean arterial pressure (MAP) in pentobarbital anesthetized rats. The left panel shows the time course of the hypotensive response to a 20 min intravenous infusion of cromakalim at three different concentrations. The average value of the baseline MAP for the three groups was 123 ± 1 mmHg. The right panel shows maximum steady state responses to cromakalim as a function of the intravenous concentration infused. Responses at all dose levels were statistically significant [Reproduced from Cavero *et al* (6) with permission.]

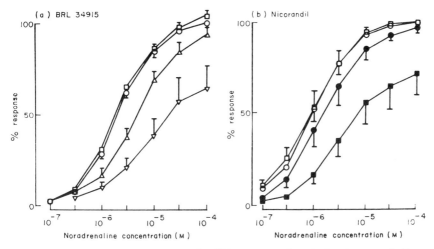

Figure 2: Effects of exposure of rabbit mesenteric arteries to cromakalim (BRL 34915) or nicorandil on subsequent responses to norepinephrine. Symbols are means and vertical bars are ± 1 SEM. Symbols refer to the following concentrations of cromakalim or nicorandil: control ($\square$), $10^{-8}$ M (O), $10^{-7}$ M ($\triangle$), $10^{-6}$ M ($\nabla$), $10^{-5}$ M ($\bullet$), and $10^{-4}$ M ($\blacksquare$). [Reproduced from Clapham and Wilson (18) with permission.]

different from that of calcium antagonists such as nifedipine or nonspecific vasodilators such as hydralazine.

### Smooth Muscle

*Effects on contractile responses.* Cromakalim and other K+ channel activators have been shown to exert an inhibitory effect on force development in smooth muscle. In spontaneously active preparations such as the rat portal vein (7,10,11,13), the guinea pig portal vein (13,14), the guinea pig trachealis (15) and taenia caeci (12), and the pig coronary artery treated with 3,4-diaminopyridine (16), cromakalim as well as nicorandil produce a dose dependent inhibition of spontaneous contractions with a threshold of about 30 nM for cromakalim and 1 µM for nicorandil. This action occurs as a result of a reduction in the frequency, the amplitude, and the duration of the spontaneous contractions.

In tonic preparations such as the rat (8,17) and rabbit thoracic aorta, and rat mesenteric artery (18,19), treatment with cromakalim, pinacidil or nicorandil attenuates responses to the subsequent administration of contractile agonists such as norepinephrine, angiotensin II and KCl in a dose dependent manner. Figure 2 shows such responses to cromakalim and nicorandil in rabbit mesenteric artery (18). Both compounds non-competitively inhibit responses to the subsequent addition of norepinephrine. Cromakalim is more effective in anatagonizing receptor mediated contractions than those produced by KCl (1,7,17,18). For the former group of agonists, cromakalim increases the $ED_{50}$ level of the agonist and non-competitively reduces the maximum response to the agonist. For KCl-mediated contractions, cromakalim antagonizes responses to low levels of KCl (ca. 20 mM) but not high levels (ca. 90 mM).

*Relaxation effects.* When arterial smooth muscle preparations such as rat (4,6,20) and rabbit thoracic aorta (8), rat mesenteric and tail artery

(20), and rabbit mesenteric artery (18,19) are contracted submaximally with an $ED_{50}$ concentration of an agonist (NE or KCl), cromakalim can inhibit this tonic force development in a dose-dependent manner with an $IC_{50}$ of about 100 nM as shown in figure 3 for rat thoracic aorta. On the other hand, as shown in figure 4, neither cromakalim nor nicorandil are able to produce significant relaxation when activation is produced by high levels of KCl (18). This is in contrast to the effects of nifedipine a known inhibitor of $Ca^{2+}$ influx through voltage-dependent channels. If the $K^+$ concentration is increased to 35 mM the relaxing effect of cromakalim against norepinephrine stimulation is abolished (8), but not the inhibitory effects of nifedipine.

**Cardiac Muscle Effects**

Several studies have been conducted to evaluate the effects of $K^+$-channel activators on cardiac tissue. Scholtysik (21) showed that cromakalim decreased the duration of the action potential in stimulated guinea pig papillary muscles partially depolarized by 22 mM $K^+$ (figure 5). Also shown in this figure for comparison are the actions of ICS 205-930 thought to be a 5-$HT_3$ receptor antagonist and sotalol, a class III antiarrhythmic agent. These latter two compounds are seen to prolong the duration of the action potential in contrast to the action of cromakalim. This action of cromakalim was interpreted to be the result of a stimulation of the $K^+$ conductance of the cell membrane.

In a paced, canine papillary muscle preparation, cromakalim injected intraarterially caused an increase in coronary blood flow, a decrease in the force of contraction and an accelerated repolarization of the action potential (22). In unpaced preparations, cromakalim decreased the rate of ventricular automaticity, decreased sinus node rate, and prolonged AV conduction time. These results were found to be similar to those produced by pinacidil and nicorandil, two other putative $K^+$ channel activators.

Kerr et al (23) studied the effects of pinacidil on experimental cardiac arrhythmias associated with coronary artery ligation in the dog. Pinacidil produced a decrease in arterial pressure and suppressed the arrhythmias present 24 hours after ligation in this animal model. This action of pinacidil was not inhibited by treatment with propranolol. Equihypotensive doses of hydralazine or sodium nitroprusside failed to exert a similar effect on the arrhythmias and verapamil treatment was similarly without effect.

Similarly, Grover et al (24) studied the cardioprotective effect of pinacidil and cromakalim on global ischemia in the isolated perfused rat heart. Pretreatment with pinacidil or cromakalim prior to the ischemic period resulted in significant improvement of reperfusion function and cardiac compliance with pinacidil being more effective in this regard.

MECHANISM OF ACTION

Studies have shown that cromakalim mediated relaxation of vascular smooth muscle does not involve the activation of conventional pathways known to mediate these effects (25). For example, it does not interact with dihydropyridine, 5-hydroxytryptamine, dopamine, $\alpha_1$-, $\alpha_2$-, or $\beta$-adrenergic receptors; it does not increase cellular levels of cAMP or cGMP; its does not inhibit phosphoinositide breakdown; and it's effects are

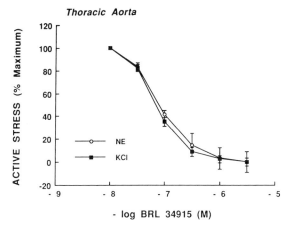

Figure 3: Relaxation responses of active force by cromakalim in rat thoracic aorta activated by $ED_{50}$ levels of norepinephrine (O) or KCl (■). Symbols are means and vertical bars ± 1 SEM. [Adapted from Cox (20) with permission.]

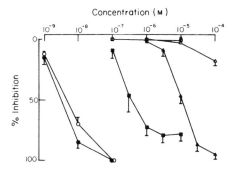

Figure 4: Concentration response curves to cromakalim (◻, ■), nicorandil (△, ▲) and nifedipine (O, ●) in rabbit mesenteric arteries. Open symbols represent tissues activated with 90 mM KCl and closed symbols tissues activated with 30 mM KCl. Vertical bars represent ± 1 SEM. [Reproduced from Clapham and Wilson (18) with permission.]

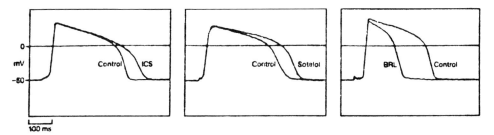

Figure 5: Effects of ICS 205-930 (30 μM), sotalol (100 μM) and cromakalim (30 μM) on action potentials in paced guinea pig papillary muscles partially depolarized by 22 mM KCl. [Reproduced from Scholtysik (21) with permission.]

31

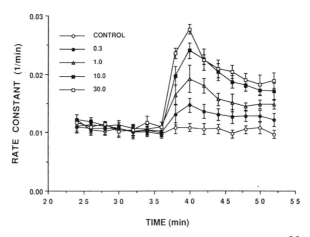

Figure 6: Concentration-response relation of cromakalim on [86]Rb efflux in rat thoracic aorta. Tissues were exposed to cromakalim (BRL-34915) during the period from 36-52 minutes. The symbols identified in the figure represent means while the vertical bars represent ± 1 SEM. [Reproduced from Cox (20) with permission.] Chromakalim values are given in μM.

not mediated through the release of endothelium-associated vasodilators. On the basis of several types of evidence, cromakalim and functionally related compounds have been suggested to act as activators of K+ channels, specifically ATP sensitive K+ channels.

## [86]Rb and [42]K Efflux Studies

Cromakalim, minoxadil and nicorandil have been shown to produce dose-dependent increases in [42]K and [86]Rb (used as a K+ marker) efflux in a variety of tissues including rat aorta and portal vein (7,8,11,13,17,20,26); guinea pig portal vein, tenia caeci and trachealis (1,12,14,26); and a variety of rabbit arteries (17,25,27).

Figure 6 shows the effects of cromakalim on the fractional rate of exchange of [86]Rb in rat thoracic aorta (20). The rate of exchange is increased by cromakalim reaching a peak effect 2-4 minutes following exposure to the compound. With continued exposure to cromakalim there is a subsequent decline in peak effect over a slower time course to a steady-state effect 15-20 minutes after the initial exposure. The effect of cromakalim is dose-dependent with the basic response being qualitatively similar at all dose levels with regard to the general time course of the response.

Some differences exist in the response of [42]K and [86]Rb efflux rates with cromakalim exposure. As shown in figure 7 (26), rates of efflux are generally higher using [42]K compared to [86]Rb under control conditions as well as during exposure to cromakalim. The responses shown in this figure from rat thoracic aorta, however, show that qualitatively similar results are obtained using the two isotopes in terms of sensitivity of cromakalim and time course of response. The results with these two isotopes are consistent with the known differences in the selectivity of K+ channels to these two ions (28).

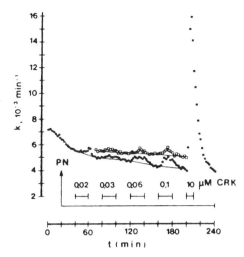

Figure 7: Effects of several concentrations of cromakalim on the rate of efflux (k) of $^{42}$K (■) and $^{86}$Rb (□) in rat thoracic aorta. Concentrations of cromakalim are given in μM. [Reproduced from Quast and Baumlin (26) with permission.]

## Electrophysiological Studies

Measurement of membrane potential by intracellular micro-electrodes in smooth muscle has shown that both cromakalim and nicorandil produce dose-dependent hyperpolarizaton toward the K+ equilibrium potential (10,15,17,29). Figure 8 shows the relationship between membrane potential and cromakalim in segments of rat portal vein (17). Significant hyperpolarizations are produced at concentrations of 0.5 μM and higher with an $IC_{50}$ value of about 1 μM. These values are similar to those obtained from $^{86}$Rb efflux responses (20).

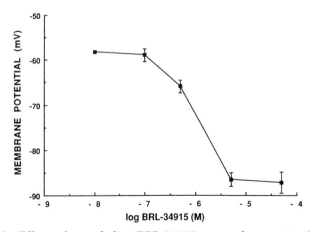

Figure 8: Effects of cromakalim (BRL 34915) on membrane potential of rat portal vein. The effects of cromakalim were determined in each cell following attaining a stable basal record of membrane potential. Symbols are means while vertical bars are ± 1 SEM. [Adapted from Weir and Weston (17) with permission].

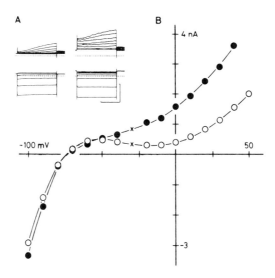

Figure 9: Effects of 10 μM pinacidil on membrane currents of isolated guinea pig ventricular myocytes in the presence of the $Ca^{2+}$ channel blocker D600 (1 μM). Panel A shows families of current tracings under control conditions (left panel pair) and with 10 μM pinacidil (right panel pair). A holding potential of -30 mV was used with depolarizing (top panels) and hyperpolarizing (bottom panels) voltage steps. The dotted lines show zero current. The calibration bars represent 150 msec and 3 nA, respectively. (B) shows the instantaneous current-voltage relation with current measured at the end of each 300 msec voltage step for control (open circles) and with pinacidil (closed circles)  The holding potential is indicated by the crosses in each current-voltage curve. [Reproduced from Iijima and Taiara (32) with permission.]

Studies made under voltage-clamp conditions also suggest that cromakalim exerts a direct action on K+ channels. Increases in outward currents have been shown using whole cell, patch clamp methods in smooth muscle cells from the rabbit portal vein (30). Cromakalim has also been shown to increase the open time of K+ channels in inside-out membrane patches of isolated rat mesenteric artery cells (31).

Another putative K+ activator, pinacidil, has been shown to effect K+ currents in single ventricular cells from the guinea pig  using whole cell patch clamp methods (32,33). As shown in figure 9, pinacidil increased inward (rectifier) K+ current and abolished the negative slope region of the current-voltage curve without effecting the time-dependent delayed outward current. These direct electrophysiological studies are the most convincing evidence that cromakalim and functionally similar compounds exert their action at the level of the K+ channel in cell membranes.

**Effects of K+ Channel Blockers**

The effects of cromakalim on $^{42}K$ (or $^{86}K$) efflux and contractions are antagonized by nonspecific K+ channel blockers such as 4-aminopyridine and tetraethylammonium ions but not by apamin or charybdtoxin, specific blockers of $Ca^{2+}$ activated K+ channels (34). However, glyburide or glibenclamide, an antidiabetic compound known to block ATP sensitive K+ channels, antagonizes the effects of cromakalim on a) $^{42}K$ (and $^{86}Rb$) efflux,

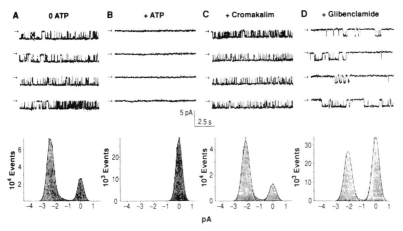

Figure 10: Effects of ATP and cromakalim on single K⁺ channel currents in a cell isolated from rabbit mesenteric arteries using an excised inside-out patch of membrane by voltage clamp procedure. The top row shows single channel records while the bottom row shows amplitude histograms of events. The membrane was bathed without ATP in (A), with 1 mM ATP in (B), with 1 mM ATP and 10 μM cromakalim in (C), and with 1 mM ATP, 10 μM cromakalim and 20 μM glibenclamide in (D). The arrows in the upper row shows the zero current level or closed channel condition. [Reproduced from Standen *et al* (31) with permission.]

b) norepinephrine and angiotensin II mediated contractions, and c) blood pressure *in vivo*. Figure 10 shows the effects of ATP, cromakalim and glibenclamide on single channel electrophysiological recordings in myocytes isolated from rat mesenteric artery. In the absence of ATP there were a significant number of single channel events associated with channel opening and closing (panel A). This activity was completely inhibited by 1 mM ATP (panel B) but was returned by the addition of 10 μM cromakalim in the continued presence of 1 mM ATP (panel C). The addition of 20 μM glibenclamide inhibited the activity induced by cromakalim in the continued presence of ATP (panel D). On the basis of these results, it has been suggested that cromakalim and related compounds exert their action through the activation of ATP-sensitive K⁺ channels. These channels are thought to be voltage insensitive but inactivated by intracellular ATP levels above 1 mM.

Arena and Kass (33) using whole cell patch clamp studies of guinea pig ventricular myocytes showed that the current activated by pinacidil was blocked by externally applied $Ba^{2+}$, $Cs^+$ and tetraethylammonium ions indicating that K⁺ ions carried this current which also showed no voltage-dependent gating. In addition, 100 nM glibenclamide potently blocked the pinacidil enhanced current. The pharmacological studies as well as the time- and voltage-independent properties of the pinacidil-sensitive current strongly suggests that pinacidil acts on the ATP-sensitive potassium channel in the heart as well.

UNRESOLVED QUESTIONS

There are several unresolved questions that require answers before activation of ATP-sensitive K⁺ channels can be accepted as the only or even primary mechanism of action of cromakalim and functionally related compounds.

## Differential Sensitivity of Actions

There is a dissociation between the concentration of cromakalim that produces inhibition of the norepinephrine induced contraction of phasic smooth muscle and the spontaneous mechanical activity of tonic smooth muscle on the one hand, and that which augments [86]Rb efflux from those tissues on the other hand (11,13,20). Figure 11 shows the effects of cromakalim on spontaneous mechanical activity and rate of efflux of [86]Rb for rat portal vein (13). A dose of 30 nM cromakalim decreased the frequency of spontaneous contractions without effecting [86]Rb efflux. Higher concentrations of cromakalim (ie, > 0.3 µM) completely inhibit spontaneous activity and also increase [86]Rb efflux.

It has been argued that cromakalim may effect a small population of (pacemaker) cells responsible for the generation of the spontaneous activity in these tissues which would not produce a measurable [86]Rb efflux response (10,14,35). However, similar observations on a dissociation between concentrations of cromakalim that inhibit force development and augment efflux have been found in tonic smooth muscle (20). Figure 12 shows a comparison of cromakalim concentration-effect relations for the inhibition of norepinephrine-induced contractile force, [86]Rb efflux, and membrane potential for rat blood vessels. The effect on contraction occurs at concentrations much less than those that effect the other two parameters. It is interesting that a very close relation exists between efflux and membrane potential suggesting a direct coupling between these two parameters with K+ channel conductance increases.

Another argument that has been made is how much of an increase in efflux is necessary to initiate relaxation of contractile force generation? Also, [86]Rb efflux is one step removed from the primary effect of these compounds on K+ channel conductance. These unanswered questions

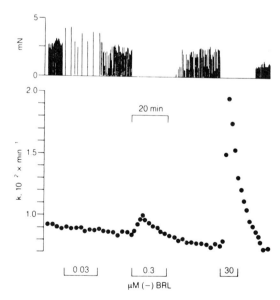

Figure 11: Effects of three different concentrations of the (-)isomer of BRL 34915 (cromakalim) on spontaneous mechanical activity (upper panel) and [86]Rb efflux rate (lower panel) measured simultaneously in rat portal vein. [Reproduced from Cook (13) with permission.]

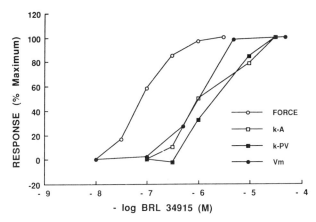

Figure 12: Comparison of the effects of cromakalim on the inhibition of norepinephrine induced contractile responses (force) and augmentation of $^{86}$Rb efflux (k - A) under basal conditions for rat thoracic aorta. Also shown for comparison are the effects of cromakalim on $^{86}$Rb efflux rate (k - PV) and membrane potential ($V_m$) in rat portal vein. [Data adapted from Cox (20), and Weir and Weston (17) with permission.]

concerning these relations may mitigate these arguments concerning cromakalim's mechanism of action.

However, there are also substantial regional differences in the magnitude of the effects of cromakalim on efflux activation and relaxation of contraction for several vascular sites in the rat. As shown in figure 13 (20), small arteries such as the tail artery and mesenteric artery branches (sites that represent models of resistance vessels) exhibit relatively small efflux responses to cromakalim whereas larger vessels such as the thoracic aorta and the portal vein exhibit large efflux responses. On the other hand, both tail arteries and mesenteric artery branches exhibit significant relaxation of norepinephrine induced contractions in response to cromakalim. Comparison of responses for efflux and relaxation in aorta and mesenteric arteries shows similar relaxation responses but substantially different efflux responses which are large in the former but only transient in nature in the latter.

The time course of the effects of cromakalim on contractions and efflux are also significantly different. At all doses, cromakalim produces an increase in efflux which reaches a peak response in 2-4 minutes after exposure. Relaxation responses on the other hand show delays before a response may begin of the order of 5-10 minutes, and may require up to 30 minutes or more to reach a steady state response level.

### Methodological Differences

Most of the experimental work performed to evaluate the effects of cromakalim on K+ channel conductance has been done using the efflux of $^{86}$Rb as marker. A lower sensitivity of $^{86}$Rb compared to $^{42}$K has been claimed to be in part responsible for the apparent dissociation between $^{86}$Rb efflux and active force. However, recent results of Quast and Baumlin (26) comparing the effect of cromakalim on $^{42}$K and $^{86}$Rb efflux suggest that the two isotopes produce qualitatively similar effects in rat thoracic aorta and guinea pig portal vein. It does not appear that the above explanation is a reasonable argument for the observed differences in effectiveness of cromakalim on efflux and force.

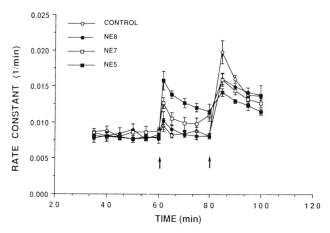

Figure 13: Comparison of the effects of cromakalim (BRL 34915) on basal $^{86}$Rb efflux (left panel) and inhibition of norepinephrine induced active force (right panel) in arteries from three arterial sites in the rat: thoracic aorta (O), mesenteric artery (■) and tail artery (Δ). Symbols represent means while vertical bars represent ± 1 SEM. [Adapted from Cox (20) with permission.]

Another problem relates to efflux responses of smooth muscle performed with protocols that are similar to those used to determine relaxation of pre-contracted tissues (20). When tissues (such as the rat thoracic aorta) are pretreated with an ED$_{50}$ dose of norepinephrine, 1 μM cromakalim produces a complete relaxation of contraction to baseline levels (figure 13). When a similar protocol is used in an efflux experiment, 10 μM cromakalim has very small effects on efflux as shown in figure 14 (20). Increasing levels of norepinephrine present before exposure of the tissue to

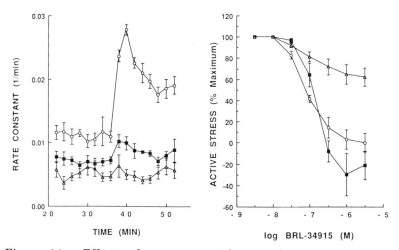

Figure 14: Effects of treatment with several concentrations of norepinephrine on the subsequent $^{86}$Rb efflux response to 10 μM cromakalim (second arrow, 80 min) in rat thoracic aorta. Symbols identify the concentration of norepinephrine used before cromakalim and represent means while vertical bars represent ± 1 SEM. [Reproduced from Cox (20) with permission.]

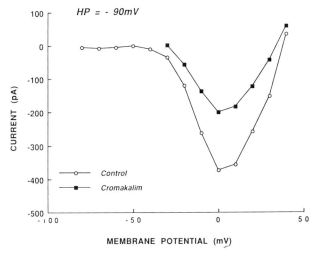

Figure 15: Current-voltage relations recorded on a freshly isolated myocyte from the rabbit portal vein at room temperature. Peak currents were recorded from a holding potential of -90 mV during 75 msec depolarizations. The internal pipet solution consisted of (in mM): 100 CsCl, 20 TEACl, 5 MgATP, 5 NaCl, 14 EGTA and 10 HEPES at pH 7.2. The external solution consisted of the following (in mM): 140 NaCl, 1 MgCl$_2$, 2 CaCl$_2$, 10 HEPES and 10 dextrose at pH 7.4. Records were obtained before and after exposure to 20 μM cromakalim. Cell capacitance was 49.7 pF.

cromakalim results in a progressive decrease in the absolute cromakalim-induced efflux response. This is difficult to reconcile with the effectiveness of cromakalim on active force with norepinephrine induced tone which shows non-competitive inhibition at all dose levels (figure 2).

## Mechanisms of Action

The primary evidence cited as proof for ATP-sensitive K$^+$ channels being the site of action of cromakalim and functionally related compounds is the ability of glyburide to antagonize the actions of cromakalim (6,8,24,31,34). In pancreatic β cells, increases in intracellular ATP block a specific population of K$^+$ channels producing membrane depolarization and insulin release (36). This action lowers plasma glucose levels *in vivo*. Glyburide blocks this channel and produces the same effects (37). However, some questions do exist concerning the action of glyburide in this regard. Glyburide inhibition of the relaxing effects of various putative K$^+$ channel activators is not constant. Winquist *et al* (33) have shown that there is a differential sensitivity in the effectiveness of several activators with glyburide being most effective against minoxadil sulfate, intermediate for cromakalim, and least effective for diazoxide induced effects.

The selectivity of the actions of glyburide (or glibenclamide) has been found to be much higher in pancreatic β cells compared to smooth muscle cells. [3H]glibenclamide has been shown to bind to pancreatic β cells with a K$_d$ of about 0.3 nM. On the other hand, glibenclamide inhibits the cromakalim activation of [86]Rb efflux rate in rat and rabbit aorta with an IC$_{50}$ of about 3 μM (8); the cromakalim-mediated relaxation of norepinephrine induced contraction in rat aorta and angiotensin II induced contractions of rabbit aorta with an IC$_{50}$ of about 0.3 μM (8); and the cromakalim-mediated relaxation of 30 mM KCl induced contractions rat aorta with an $IC_{50}$ of about 0.1 μM (6).

Cavero *et al* (6) have shown that intravenous bolus injection of glibenclamide inhibits the hypotensive effects of cromakalim. They have also shown that the intravenous doses of glibenclamide required for this effect are 100-300 times larger than that required to increase insulin secretion and decrease plasma glucose *in vivo*. These findings are consistent with the above described selectivity difference in the effects of glibenclamide on pancreatic β versus smooth muscle cells.

It has been demonstrated *in vivo* that intravenous injections of glibenclamide alone produce very small pressor responses and that *in vitro* doses that completely inhibit the relaxation effects of cromakalim on smooth muscle contraction have no effect themselves of active tone of smooth muscle *in vitro* (6). This has been interpreted to suggest that ATP-sensitive $K^+$ channels do not participate in the regulation of arterial pressure under normal conditions. This further suggests that these channels are already inactivated or blocked by intracellular levels of ATP under normal conditions. How then does cromakalim activate these channels if they are normally inactivated by ATP *in vivo*? This might suggest that ATP exerts its effect not by actual channel block but by modulation (inhibition) of the open conductance state of the channel through an interaction with a specific, high sensitivity binding site. Then cromakalim must have the ability to overcome this inhibition of these channels by an independent mechanism that causes an increase in the open probability of channels. However, high affinity binding sites for cromakalim have not been identified in smooth muscle cells (25). Thus, the exact molecular mechanism by which cromakalim activates $K^+$ channels remains undetermined. The elucidation of this mechanism becomes an important prerequisite for understanding the actions of these agents.

In preliminary experiments by the author, cromakalim has been found to exert a direct effect on membrane currents in single, enzymatically dispersed cells of the rabbit portal vein measured under whole cell, voltage clamp conditions in which $K^+$ currents were inhibited. Figure 15 shows current-voltage relations recorded in the same cell with 100 mM CsCl and 20 mM TEA-Cl in the patch pipet to block $K^+$ currents, no KCl added to the external or pipet solutions, and 2 mM $Ca^{2+}$ in the external solution. In the presence of 20 μM cromakalim, these inward currents were inhibited at every value of membrane potential compared to the control. This suggests that cromakalim either inhibits an inward current or augments an outward current in this tissue. However, it is not clear which ion(s) would carry such an outward current under the conditions employed. This finding needs to be confirmed and explored, however.

None of these questions or problems are necessarily fatal with regard to the hypothesis concerning the mechanism of action of cromakalim. They can be interpreted to suggest that one of the following is correct: a) the relationship between the amount of an increase in $K^+$ channel conductance and the regulation of contractile protein interaction is not fully understood; b) the effects of intracellular (compensatory) mechanisms regulating $K^+$ channel conductance is not fully understood; c) differences exist in tissue selectivity for $K^+$ channel activators and their inhibitors; and/or d) more than one mechanism exists for the effects of cromakalim and related compounds *in vivo* and *in vitro*.

## ACKNOWLEDGMENT

The author wishes to express his gratitude to Drs. Ronald Ferrone and George Oshiro for their stimulation of his interest in this class of compounds, to Ms. Elaine Veit and Sue Van Buren for their tireless assistance in these studies, and to Ms. Maxine Blob for her secretarial and editorial assistance with this manuscript. The cromakalim was generously supplied by Dr. T.C. Hamilton of Beechem Pharmaceuticals. The research in the author's laboratory was supported by research grants HL28476, HL39388 and AG04908 from the USPHS, and by Grant-in-Aid 85 1143 from the American Heart Association.

## REFERENCES

1. Cook NS, Quast U, Hof RP, Baumlin Y, Pally C. Similarities in the mechanisms of action of two new vasodilator drugs: pinacidil and BRL 34915. *J Cardiovasc Pharmacol* 11: 90, 1988.
2. Jones AW. Content and fluxes of electrolytes. *Handbook of Physiology, The Cardiovascular System, VII, Vascular Smooth Muscle*, Bethesda: American Physiological Society, 1980, p 253.
3. Stekiel W. Electrophysiological mechanisms of force development by vascular smooth muscle membrane in hypertension. In: *Blood Vessel Changes in Hypertension: Structure and Function, VII*, R.M.K.W. Lee (ed). Boca Raton: CRC Press, pp 127, 1989.
4. Buckingham RE, Clapham JC, Hamilton TC, Longman SD, Nortaon J, Poyser RH. BRL 34915, a novel antihypertensive agent: Comparison of effects on blood pressure and other hemodynamic parameters with those of nifedipine in animal models. *J Cardiovasc Pharmacol* 8: 798, 1986.
5. Buckingham RE, Hamilton TC, Howlett DR, Mootoo S, Wilson C. Inhibition by glibenclamide of the vasorelaxant action of cromakalim in the rat. *Br J Pharmacol* 97: 57, 1989.
6. Cavero I, Mondot S, Mestre M. Vasorelaxant effects of cromakalim in rats are mediated by glibenclamide-sensitive potassium channels. *J Pharmacol Exp Ther* 248: 1261, 1989.
7. Hof RP, Quast U, Cook NS, Blarer S. Mechanism of action and systemic and regional hemodynamics of the potassium channel activator BRL34915 and its enantiomers. *Circ Res* 62: 679, 1988.
8. Quast U, Cook NS. In vitro and in vivo comparison of two K$^+$ channel openers, diazoxide and cromakalim, and their inhibition by glibenclamide. *J Pharmacol Exp Ther* 250: 261, 1989.
9. VandenBerg MJ, Woodward SR, Hossain M, Stewart-Long P, Tasker TCG. Potassium channel activators lower blood pressure: an initial study of BRL 34915 in hypertensive patients. *J Hypertension* 4: S166, 1986
10. Hamilton TC, Weir SW, Weston AH. Comparison of the effects of BRL 34915 and verapamil on electrical and mechanical activity in rat portal vein. *Br J Pharmacol* 88: 103, 1986.
11. Shetty SS, Weiss GB. Dissociation of actions of BRL 34915 in the rat portal vein. *Eur J Pharmacol* 141: 485, 1987.
12. Weir SW, Weston AH. Effect of apamin on responses to BRL 34925, nicorandil and other relaxants in the guinea-pig taenia caeci. *Br J Pharamcol* 88:113, 1986.
13. Cook NS. The pharmacology of potassium channels and their therapeutic potential. *TIPS* 9: 21, 1988.

14. Quast U. Effects of the K+ efflux stimulating vasodilator BRL 34915 on 86Rb+ efflux and spontaneous activity in guinea-pig portal vein. *Br J Pharmacol* 91: 569, 1987.
15. Allen SL, Boyle JP, Cortijo J, Foster RW, Morgan GP, Small RC. Electrical and mechanical effects of BRL34915 in guinea-pig isolated trachealis. *Br J Pharmacol* 89: 395, 1986.
16. Kuromaru O, Sakai K. Effects of nicorandil on the rhythmic contractile response of isolated miniature pig coronary artery to 3,4-diaminopyridine: Comparison with nitroglycerin. *Arch Int Pharmacol Therap* 283: 272, 1986.
17. Weir SW, Weston AH. The effects of BRL 34915 and nicorandil on electrical and mechanical activity and on 86Rb efflux in rat blood vessels. *Br J Pharmacol* 88: 121, 1986.
18. Clapham JC, Wilson C. Anti-spasmogenic and spasmolytic effects of BRL 34915: a comparison with nifedipine and nicorandil. *J Auton Pharmacol* 7: 233, 1987.
19. Wilson C, Coldwell MC, Howlett DR, Cooper SM, Hamilton TC. Comparative effects of K+ channel blockade on the vasorelaxant activity of cromakalim, pinacidil and nicorandil. *Eur J Pharmacol* 152: 331, 1988.
20. Cox RH. Effects of putative K+ channel activator BRL-34915 on arterial contraction and 86Rb efflux. *J Pharmacol Exp Ther* 252:51, 1990.
21. Scholtysik G. Evidence for inhibition by ICS 205-930 and stimulation by BRL 34915 of K+ conductance in cardiac muscle. *Naun-Schmied Arch Pharmacol* 335: 692, 1987.
22. Gotanda K, Satoh K, Taira N. Is the cardiovascular profile of BRL 34915 characteristic of potassium channel activators? *J Cardiovasc Pharmacol* 12: 239, 1988.
23. Kerr MJ, Wilson R, Shanks RG. Suppression of ventricular arrhythmias after coronary artery ligation by pinacidil, a vasodilator drug. *J Cardiovasc Pharmacol* 7: 875, 1985.
24. Grover GJ, McCullough JR, Henry DE, Condor ML, Sleph PG. Anti-ischemic effects of the potassium channel activators pinacidil and cromakalim, and the reversal of these effects with the potassium channel blocker glyburide. *J Pharmacol Exp Ther* 251: 98, 1989.
25. Coldwell MC, Howlett DR. Specificity of action of the novel antihypertensive agent, BRL 34915, as a potassium channel activator. *Biochem Pharmacol* 36: 3663, 1987.
26. Quast U, Baumlin Y. Comparison of the effluxes of 42K+ and 86Rb+ elicited by cromakalim (BRL 34915) in tonic and phasic vascular tissue. *Naun-Schmied Arch Pharmacol* 338: 319, 1988.
27. Post JM, Smith JM, Jones AW. BRL 34925 (cromakalim) stimulation of 42K efflux from rabbit arteries is modulated by calcium *J Pharmacol Exp Ther* 250: 591, 1989.
28. Latorre R, Miller C. Conduction and selectivity in potassium channels. *J Membr Biol* 71: 11, 1983.
29. Yamanaka K, Furukawa K, Kitamura K. The different mechanisms of action of nicorandil and adenosine triphosphate on potassium channels of circular smooth muscle of the guinea-pig small intestine. *Naun-Schmied Arch Pharmacol* 331: 96, 1985.
30. Beech DJ, TB Bolton. Effects of BRL 34915 on membrane currents recorded from single smooth muscle cells from the rabbit portal vein. *Br J Pharmacol* 92: 550P, 1987.

31. Standen NB, Quayle JM, Davies NW, Brayden JE, Huang Y, Nelson MT. Hyperpolarizing vasodilators activate ATP-sensitive K+ channels in arterial smooth muscle. *Science* 245: 177, 1989.
32. Iijima T, Taira N. Pinacidil increases the background potassium current in single ventricular cells. *Eur J Pharmacol* 141: 139, 1987.
33. Arena JP, Kass RS. Enhancement of potassium-sensitive current in heart cell by pinacidil. Evidence for modulation of the ATP-sensitive potassium channel. *Circ Res* 65: 4366, 1989.
34. Winquist RJ, Henry LA, Wallace AA, Baskin EP, Stein RB, Garcia ML, Kaczorowski GJ. Glyburide blocks the relaxation response to BRL 34915 (cromakalim), minoxidil sulfate and diazoxide in vascular smooth muscle. *J Pharmacol Exp Ther* 248: 149, 1989.
35. Hollingsworth M, Amedee T, Edwards D, Mironneau J, Savineau JP, Small RC, Weston AH. The relaxant effect of BRL 34915 in rat uterus. *Br J Pharmacol* 91: 803, 1987.
36. Cook DL, CN Hales CN. Intracellular ATP directly blocks K+ channels in pancreatic cells. *Nature* (Lond) 311: 271, 1984.
37. Sturgess NC, Ashford MLJ, Cook DL, Hales CN. The sulphonylurea receptor may be an ATP-sensitive potassium channel. *Lancet* 2: 474, 1985.

# EFFECTS OF HYPERTENSION ON ARTERIAL GENE

# EXPRESSION AND ATHEROSCLEROSIS

Aram V. Chobanian

Boston University School of Medicine
Boston, MA 02118

## INTRODUCTION

Hypertension may induce major changes in the arterial intima and media which may lead to arterial injury and atherosclerosis. Although the mechanisms involved in these alterations have not been clearly defined, recent studies in our own and other laboratories have demonstrated changes in gene expression of certain growth factors and functionally important proteins in response to blood pressure elevation or to vasoactive agents. This paper reviews these observations and also summarizes some recent data on the effects of antihypertensive drugs on atherogenesis.

## HYPERTENSION AND GROWTH OF VASCULAR CELLS

Hypertension causes an increase in the cellular mass of the arterial media, primarily by inducing hypertrophy of medial smooth muscle cells (SMC). Recent studies involving cultured SMC have suggested that vasoactive hormones may influence the growth of these cells. Angiotensin II has been shown to increase size and protein content of SMC in culture (1). The hypertrophy of these cells in response to angiotensin II appears to be associated with increased steady-state mRNA levels of the protooncogenes c-myc (2) and c-jun (3) and of the A chain of platelet-derived growth factor (PDGF) (2). Other vasoconstrictor agents including catecholamines (4), endothelin (5) and serotonin (6) also stimulate growth of cultured SMC.

We recently have been involved in the examination of the gene expression of arterial growth factors in vivo in rat aorta and the influence of hypertension on such expression. We have used seven cDNA probes complementary to regions of PDGF A and B chains, to insulin growth factors (IGF) I and II, to endothelial cell growth factor (ECGF), to basic fibroblast growth factor (bFGF) and to transcription growth factor-ß (TGF-ß). Using Northern blotting techniques, we demonstrated that each of these growth factors is transcriptionally active in vivo. Hypertension induced by administration of deoxycorticosterone and salt (DOC/salt) increased the expression of TGF-ß though not of the other six growth factors (7).

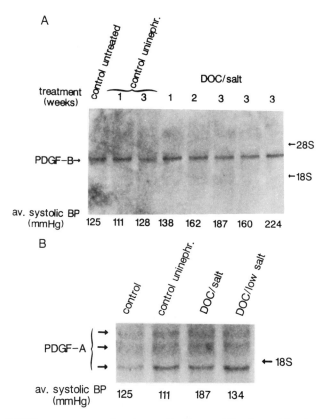

Figure 1. PDGF gene expression in aortic tissue. Northern blot hybridiza. n analysis of rat aortic RNA from hypertensive and normotensive animals. (A) PDGF B chain gene expression. Weeks of treatment were as indicated above each lane. Control untreated and uninephrectomized animals were age-matched. 1 d exposure time. (B) PDGF A chain gene expression. 2 d exposure time. Taken from reference 7.

Representative data from these studies are shown in Figures 1 and 2. The findings demonstrate that several growth factors are made in aortic tissue and could cause autocrine or paracrine effects. Since TGF-ß may be involved in several important functions that may be associated with vascular injury such as wound healing and modulation of cell growth and of connective tissue synthesis (8), the increase in expression of TGF-ß with DOC-salt hypertension could have important functional significance in arterial tissue. Owens and associates recently have observed that TGF-ß causes hypertrophy of SMC in culture and increases the number of polyploid cells (9). These findings are of particular interest since chronic hypertension in the rat can induce a marked increase in the tetraploid population of aortic SMC (10-12) as a result of DNA endoreplication but without cell division.

We have been interested in the effects of hypertension on fatty acid binding proteins (FABP) that represent a multigene family of intracellular proteins present in high concentrations in the cytosol (13,14). These proteins bind several organic substances including fatty acids and retinol and may be involved in the regulation of cell growth and differentiation (15,16). We demonstrated, using polyclonal antibodies, the presence in the rat aorta of the type of FABP found in rat heart (hFABP) (17). We also found

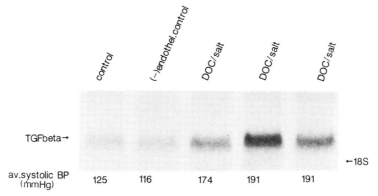

Figure 2. TGF-ß gene expression in aortic tissue. Northern blot hybridization analysis of rat aortic RNA from normotensive and hypertensive animals. 20 µg of total RNA from different groups of rat aortas was applied to each lane. 18S indicates the position of ribosomal RNA. Average systolic blood pressure from each group of animals is shown at the bottom of each lane. (-) endothel. control, RNA extracted from endothelium-scraped aortas of untreated rats. DOC/salt groups were all treated for 3 wk 2 d exposure time. Adapted from reference 7.

that the steady-state mRNA level for hFABP in aorta was markedly reduced by DOC/salt hypertension (Figure 3) as was the amount of immunologically detectable protein. This effect appeared related to the blood pressure increase since it was not present in other tissues and was also observed in the aortas of two-kidney, one-clip hypertensive rats and of rats infused with angiotensin II for 6 days (Figure 4). How such changes in hFABP affect the metabolism or growth of vascular cells is unknown as yet, but the problem needs to be addressed in future studies, particularly since hFABP normally is present in such high concentrations in vascular tissue.

NA,K-ATPASE

Considerable interest has centered on the potential role of the sodium pump in the pathogenesis or maintenance of hypertension. Several isoforms of the enzyme have previously been identified (18), and Herrera and associates at our institution have examined the changes in expression of subunit isoforms of Na,K-ATPase in response to hypertension (19). Using Northern blot analysis (Figure 5), we observed that the expression of the alpha-1 subunit increased considerably in aorta of DOC/salt hypertensive rats. Increases also were seen in skeletal muscle, suggesting that the alterations were not related to changes in blood pressure but were attributable to some other action of deoxycorticosterone and/or salt. On the other hand, with the alpha-2 subunit, there was a marked reduction in expression in aorta and heart but not in tissues that are not directly exposed to the effects of elevated blood pressure, such as skeletal muscle. Reduction in expression of the alpha-2 subunit in aorta also was induced by 1 week of angiotensin II infusion, suggesting again that the changes were related to blood pressure elevation. The functional consequences of such alterations and their importance in the hypertensive process remain to be determined. However, of interest has been the recent observation from Herrera's laboratory demonstrating a genetic polymorphism in the alpha-2 subunit in Dahl salt-sensitive rats but not in their salt-resistant controls (20).

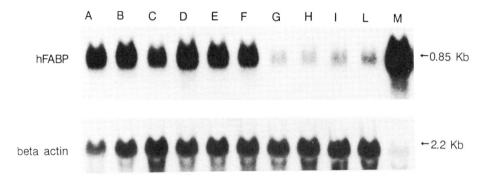

Figure 3. Effect of DOC/salt hypertension on hFABP expression in aortic tissue by Northern blot analysis. Each lane contains 20 µg of total RNA. Lanes A,B,C, uninephrectomized rats untreated for 1, 4 and 5 weeks, respectively; lanes D-L, uninephrectomized rats treated with DOC/salt for 3 days (lane D); 1 week (lane E); 2 weeks (lane F); 3 weeks (lane G); and 4 weeks (lanes H,I,L). Lane M, heart total RNA from an untreated control rat. Upper, rat hFABP cDNA used as the labeled probe. Lower, the same blot hybridized with a rat actin cDNA probe. Adapted from reference 17.

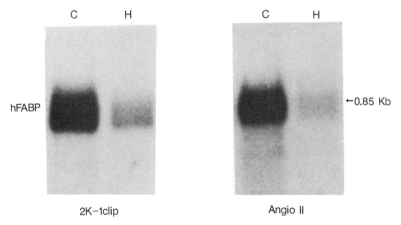

Figure 4. Effect of Goldblatt two-kidney, one-clip renal hypertension (2K-1 clip) and angiotensin II infusion (Angio II) on expression of rat hFABP in aortic tissue. Each lane contains 20 µg of total RNA from a pool of four aortas from sham-treated controls (C) or hypertensive animals (H) following treatment for 4 weeks (2K-1 clip model) or 6 days (Angio II model). Adapted from reference 17.

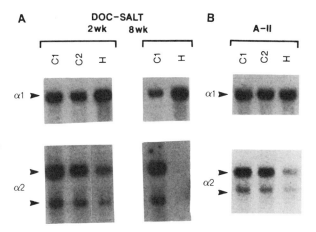

Figure 5. Na$^+$,K$^+$-ATPase alpha subunit mRNAs in aortas from hypertensive rats. (A) mRNA of Na$^+$,K$^+$-ATPase subunits alpha-1 and alpha-2 were analyzed in separate RNA blots with equivalent amounts of total cellular RNA derived from pooled aortas from hypertensive, uninephrectomized rats treated with DOC/salt (H), control uninephrectomized rats (C1), and control normotensive DOC-low salt rats (C2). The specific hybridizing bands to the respective probes are indicated by arrowheads. Two hybridizing bands are detected in alpha-2, representing two sizes of mRNAs. (B) Na$^+$,K$^+$-ATPase alpha subunit mRNAs, alpha-1 and alpha-2, were analyzed in aortic total cellular RNA from rats made hypertensive by 1-week intraperitoneal A-II infusion. C1, control normotensive rats with pump implanted to infuse medium. C2, control nomotensive rats with subpressor dose of A-II infused (75 ng/min). H, hypertensive rats with pressor dose of A-II infused (200 ng/min). Adapted from reference 19.

## HYPERTENSION AND ATHEROSCLEROSIS

Hypertension causes many changes in the arterial intima. The endothelium is altered in appearance (21,22) and there is increased permeability of the endothelium to a variety of substances including macromolecules (23-25). In addition, endothelium-derived relaxation of arterial tissue appears to be inhibited in several hypertensive models (26). Hypertension also somehow enhances the adherence of white blood cells, including monocytes, to the endothelial surface and promotes their entry and accumulation in the subendothelial space (21). Despite these changes, which also can be observed in the development of the atherosclerotic lesion, hypertension by itself does not appear to cause atherosclerosis. On the other hand, in the presence of hypercholesterolemia, we believe that hypertension is a potent promoter of atherosclerosis.

To test the effects of hypertension in hypercholesterolemic rabbits, we have examined the influence of one-kidney, one-clip hypertension on the course of aortic atherosclerosis in the Watanabe heritable hyperlipidemic rabbit (WHHL) which has a genetic defect in the cellular receptor for low density lipoproteins. Periods of hypertension as brief as 3 months or less in duration were associated with a marked aggravation in aortic atherosclerosis (27). Both the extent of aortic surface involvement (Table 1) and aortic cholesterol content were markedly increased in the hypertensive as compared with the normotensive WHHL.

## Table 1
### Effects of One-Kidney, One-Clip Goldblatt Hypertension on Aortic Surface Atherosclerosis in Watanabe Heritable Hyperlipidemic Rabbits

| Aortic Region | Control | Hypertensive |
|---|---|---|
| | (% intimal surface) | |
| Total Aorta | $16 \pm 3$ | $77 \pm 4$* |
| Ascending & arch | $47 \pm 15$ | $99 \pm 1$+ |
| Descending thoracic | $6 \pm 1$ | $89 \pm 5$* |
| Abdominal | $15 \pm 2$ | $37 \pm 11$ |

* $p < 0.001$. Values represent the mean $\pm$ S.E. + $p = 0.2$. Adapted from reference 27.

## EFFECTS OF ANTIHYPERTENSIVE DRUGS ON THE ARTERIAL WALL

Certain antihypertensive agents, particularly beta adrenergic blockers and calcium antagonists, have been shown to reduce the rate of development of atherosclerosis in cholesterol-fed normotensive animals (28-31). The mechanism for the anti-atherogenic effect of any of these antihypertensive drugs is unknown although all are obviously capable of lowering blood pressure or may otherwise change the hemodynamic stresses on the arterial wall.

Despite their favorable effects on atherogenesis in the cholesterol-fed animal models, beta blockers and calcium antagonists have failed as yet to affect atherosclerosis in the WHHL. Studies with propranolol (32), nifedipine (33), and verapamil (34) have proven to be negative in this regard.

Because of our interest in the growth of vascular SMC and the potential relationship of angiotensin II to such growth, we decided to study the effects of the angiotensin converting enzyme inhibitor, captopril, on aortic atherosclerosis in the WHHL35. Captopril was administered in the diet in doses of 25-50 mg/kg body weight/day from 3 to 12 months of life. A significant reduction in aortic atherosclerosis was observed in the captopril-treated as compared with control WHHL, the major inhibition occurring in the descending thoracic aorta (Table 2).

## Table 2
### Effects of Captopril on Aortic Surface Atherosclerosis in Watanabe Heritable Hyperlipidemic Rabbits

| Aortic Region | Control | Hypertensive |
|---|---|---|
| | (% intimal surface) | |
| Total Aorta | $47.9 \pm 3.6$ | $29.7 \pm 3.9$* |
| Ascending & arch | $76.3 \pm 4.3$ | $76.3 \pm 3.7$ |
| Descending thoracic | $48.9 \pm 5.2$ | $15.3 \pm 3.9$** |
| Abdominal | $28.6 \pm 2.5$ | $26.2 \pm 4.4$ |

* $p < 0.01$. Values represent the mean $\pm$ S.E. ** $p < 0.001$. Adapted from reference 35.

The anti-atherosclerotic effect was independent of reductions in plasma cholesterol but was associated with significant decreases in blood pressure in the captopril-treated animals. The effect of treatment on the morphology of the lesions also was of interest. At any given level of lesion severity, as judged by gross inspection, the atherosclerotic plaques in the captopril-treated WHHL were less cellular and had relatively greater connective tissue content than plaques in the control WHHL. The thickness of the intima and media also appeared to be reduced by captopril. The findings could in part be due to an effect of captopril on cellular growth. Support of such a mechanism has been provided by recent observations demonstrating that cilazapril, another angiotensin converting enzyme inhibitor, and captopril both caused a reduction in myointimal thickening following balloon-catheter injury of rat carotid artery (36). However, captopril may have multiple other actions as well and its mechanism of action in the WHHL remains to be determined.

## SUMMARY AND CONCLUSIONS

Hypertension and antihypertensive drugs can lead to a number of structural and functional alterations in the arterial wall. The recent studies that have been summarized in this paper have emphasized primarily the work being performed in my own laboratory, and have not attempted to review thoroughly the exciting research now taking place in this field. However, they do illustrate the fact that studies involving vascular injury and atherosclerosis have begun to change in focus to emphasize the cellular and molecular mechanisms that are responsible for the pathologic changes that develop. Such investigations ultimately should lead to improved methods for the prevention or treatment of the vascular consequences of hypertension.

Supported by NIH Grant HL18318

## REFERENCES

1. Geisterfer AAT, Peach MJ, Owens G. Angiotensin II induces hypertrophy, not hyperplasia, of cultured rat aortic smooth muscle cells. *Circ Res* 62:749-756, 1988.

2. Naftilan AJ, Pratt RE, Dzau VJ. Induction of platelet-derived growth factor A-chain and c-myc gene expressions by angiotensin II in cultured rat vascular smooth muscle cells. *J Clin Invest* 83:1419-1424, 1989.

3. Naftilan AJ, Gilliland GK, Eldridge CS, Karin M, Kraft AS. Induction of the protooncogene c-jun by angiotensin II. *Circulation* 80: II-459a, 1989.

4. Yamori Y, Mano M, Nara Y, Horie R. Catecholamine-induced polyploidization in vascular smooth muscle cells. *Circulation* 75: I-92-I-95, 1987.

5. Komuro I, Kurihara H, Sugiyama T, Takaku F, Yazaki Y. Endothelin stimulates c-fos and c-myc expression and proliferation of vascular smooth muscle cells. *FEBS Letters* 238: 249-252, 1988.

6. Nemecek GM, Coughlin SR, Handley DA, Moskowitz MA. Stimulation of aortic smooth muscle cell mitogenesis by serotonin. *Proc Natl Acad Sci USA* 83: 674-678, 1986.

7.  Sarzani R, Brecher P, Chobanian AV. Growth factor expression in aorta of normotensive and hypertensive rats. *J Clin Invest* 83: 1404-1408, 1989.

8.  Sporn MB, Roberts AB. Transforming growth factor-B. Multiple actions and potential clinical applications. *JAMA* 262: 938-941, 1989.

9.  Owens GK, Geisterfer AT, Yang YW, Komoriya A. Transforming growth factor-B-induced growth inhibition and cellular hypertrophy in cultured vascular smooth muscle cells. *J Cell Biol* 107: 771-780, 1988.

10. Owens GK, Schwartz SM. Alterations in vascular smooth muscle mass in the spontaneous hypertensive rat. Role in cellular hypertrophy, hyperploidy and hyperplasia. *Circ Res* 51: 280-289, 1982.

11. Owens GK, Schwartz SM. Vascular smooth muscle cell hypertrophy and hyperploidy in the Goldblatt hypertensive rat. *Circ Res* 53: 491-501, 1983.

12. Lichtenstein AH, Brecher PI, Chobanian AV. Effect of DOC/salt hypertension on cell ploidy in the rat aorta. *Hypertension* 8: II-50-II-54, 1986.

13. Offner GD, Troxler RF, Brecher P. Characterization of a fatty acid binding protein from rat heart. *J Biol Chem* 261: 5584-5589, 1986.

14. Sweetser DA, Heuckeroth RO, Gordon JI. The metabolic significance of mammalian fatty-acid-binding proteins: Abundant proteins in search of a function. *Annu Rev Nutr* 7: 337-359, 1987.

15. Bassuk JA, Tsichlis PN, Sorof S. Liver fatty acid binding protein is the mitosis-associated polypeptide target of a carcinogen in rat hepatocytes. *Proc Natl Acad Sci (USA)* 84: 7547-7551, 1987.

16. Distel RJ, Ro H-S, Rosen BS, Groves DL, Spiegelman BM. Nucleoprotein complexes that regulate gene expression in adipocyte differentiation: Direct participation of *c-fos*. *Cell* 49: 835-844, 1987.

17. Sarzani R, Claffey KP, Chobanian AV, Brecher P. Hypertension induces tissue-specific gene suppression of a fatty acid binding protein in rat aorta. *Proc Natl Acad Sci (USA)* 85: 7777-7781, 1988.

18. Shull GE, Greeb J, Lingrel JB. Molecular cloning of three distinct forms of the $Na^+,K^+$-ATPase alpha-subunit from rat brain. *Biochemistry* 25: 8125-8129, 1986.

19. Herrera VL, Chobanian AV, Ruiz-Opazo N. Isoform-specific modulation of Na,K-ATPase alpha-subunit gene expression in hypertension. *Science* 241: 221-223, 1988.

20. Herrera VL, Ruiz-Opazo N. Identical Na,K-ATPase alpha-1 gene polymorphism in two genetic rat strains. *Hypertension* 12: 338a, 1988.

21. Haudenschild CC, Prescott MF, Chobanian AV. Effects of hypertension and its reversal on aortic intimal lesions of the rat. *Hypertension* 2: 33-44, 1980.

22. Chobanian AV, Prescott MF, Haudenschild CC. Aortic endothelial changes during the development and reversal of experimental hypertension. In: Gotto A, Smith LC, (eds): *Atherosclerosis V.* New York: Springer-Verlag, 1980, pp 699-702.

23. Wiener J, Lattes RG, Meltzer BG, Spiro D. The cellular pathology of experimental hypertension: IV. Evidence for increased vascular permeability. *Am J Pathol* 54: 187-207, 1969.

24. Huttner I, Boutet M, Rona G, More RH. Studies of protein passage through arterial endothelium: III. Effect of blood pressure levels on the passage of fine structure protein tracers through rat arterial endothelium. *Lab Invest* 29: 536-546, 1973.

25. Chobanian AV, Brecher PI, Haudenschild CC.   Effects of hypertension and antihypertensive therapy on atherosclerosis. *Hypertension* 8: I-15-I-21, 1986.
26. Vanhoutte, PM.   Endothelium and control of vascular function. *Hypertension* 13: 658-667, 1989.
27. Chobanian AV, Lichtenstein AH, Nilakhe V, Haudenschild CC, Drago R, Nickerson C.   Influence of hypertension on aortic atherosclerosis in the Watanabe rabbit. *Hypertension* 14: 203-209, 1989.
28. Chobanian AV, Brecher P, Chan C.   Effects of propranolol on atherogenesis in the cholesterol-fed rabbit. *Circ Res* 56: 755-762, 1985.
29. Kaplan JR, Manuck SB, Adams MR, Weingand KW, Clarkson TB. Inhibition of coronary atherosclerosis by propranolol in behaviorally predisposed monkeys fed an atherogenic diet. *Circulation* 76: 1364-1372, 1987.
30. Henry PD, Bentley KI. Suppression of atherosclerosis in cholesterol-fed rabbits treated with nifedipine. *J Clin Invest* 68: 1366-1369, 1981.
31. Chobanian AV.   Effects of calcium channel antagonists and other antihypertensive drugs on atherogenesis. *J Hypert*  5: S43-S48, 1987.
32. Lichtenstein AH, Drago R, Nickerson C, Prescott MF, Lee SQ, Chobanian AV.   Effect of propranolol on atherogenesis in the Watanabe heritable hyperlipidemic rabbit.  *J Vasc Med Biol* 1: 248-254, 1989.
33. Van Niekerk JLM, Hendriks Th, DeBoer HHM, Van't Laar A.  Does nifedipine suppress atherogenesis in WHHL rabbits? *Atherosclerosis* 53: 91-98, 1984.
34. Tilton GD, Buja LM, Bilheimer DW, Apprill P, Ashton J, McNatt J, Kita T, Willerson JT.  Failure of a slow channel calcium antagonist, verapamil, to retard atherosclerosis in the Watanabe heritable hyperlipidemic rabbit:   an animal model of familial hyper-cholesterolemia. *J Amer Coll Cardiol* 6:141-144, 1985.
35. Chobanian AV, Haudenschild CC, Nickerson C, Drago R.  Anti-atherogenic effect of captopril in the Watanabe heritable hyperlipidemic rabbit. *Hypertension*, 15:327-331, 1990.
36. Powell JS, Clozel J-P, Muller RKM, Kuhn H, Hefti F, Hosang M, Baumgartner HR.   Inhibitors of angiotensin-converting enzyme prevent myointimal proliferation after vascular injury. *Science* 245: 186-188, 1989.

# ALTERED PHOSPHOLIPASE ACTIVITIES RELATED TO $\alpha_1$-ADRENERGIC RECEPTOR SUPERSENSITIVITY OF AORTAS FROM ALDOSTERONE-SALT HYPERTENSIVE RATS

Allan W. Jones, Shivendra D. Shukla, Brinda B. Geisbuhler, Susan B. Jones & Jacquelyn M. Smith*

Departments of Physiology and Pharmacology, University of Missouri
Columbia, MO 65212
*Current Address: Department of Physiology, Chicago College of
Osteopathic Medicine, Downers Grove, IL 60515

## INTRODUCTION

Vascular smooth muscle from several hypertensive models exhibits supersensitivity to catecholamines, which precedes the development of overt hypertension in rats given mineralocorticoid-salt treatment (1-5). In previous studies we did not find significant changes in $\alpha_1$-receptor characteristics (receptor type, dissociation constant or maximal receptor binding) in aorta from aldosterone-salt hypertensive rats (AHR) (6,7). Since we observed a 2-15 fold leftward shift in the NE $EC_{50}$ (concentration of norepinephrine required for 50% response) for functional responses in AHR, we concluded that increased efficacy of post receptor events occurs in AHR (7,8).

The production of second messengers (inositoltrisphosphate, $IP_3$, and diacylglycerol, DAG) from phospholipase C (PLC) catalyzed hydrolysis of phosphatidylinositol 4,5-bisphosphate ($PIP_2$) during receptor occupancy is thought to stimulate Ca release and influx (8,9). NE increased the production of inositol phosphates in rat aorta. Moreover, a significant increase was observed in AHR and stroke prone spontaneously hypertensive rats (SHRSP) (8,10). The shifts to a lower $EC_{50}$ were consistent with data from Ca-dependent functional responses which led to the conclusion that supersensitivity results from altered coupling between receptor occupancy and PLC activity. The Ca sensitivity of the contractile response in skinned aorta from AHR was similar to that in controls which indicates that the supersensitivity may result more from altered Ca regulatory mechanisms (e.g. production of second messengers) than altered sensitivity of the contractile apparatus to Ca (11).

Although the recent focus on mechanisms of receptor supersensitivity in vascular smooth muscle (VSM) has centered predominantly on the regulation of PLC activity, this enzyme may not be unique in the phospholipid signaling process. Recent studies indicate the presence of phospholipase D (PLD) activity in mammalian cells (12-14). PLD catalyzes the formation of phosphatidic acid (PA) directly from phospholipids such as phosphatidylcholine (12). DAG can then be formed from PA by the action of phosphatidic acid-phosphohydrolase (PA phosphatase) (15). PLD also catalyzes a transphosphatidylation reaction such as that between choline on phosphatidylcholine and ethanol thus forming phosphatidylethanol (PEth) (12). Since tissues from rats are not normally exposed to ethanol the endogenous PEth levels are minimal. Thus, agonist-stimulated production of PEth in tissues which are acutely exposed to ethanol provides a means to identify PLD activity. Currently, definitive information is lacking concerning the presence of agonist-stimulated PLD activity in VSM.

Increased production of arachidonic acid metabolites also occurs during NE stimulation with the prostacyclin (PGI$_2$) metabolite, 6-keto-PGF$_{1\alpha}$, being a major product (16-19). The production of 6-keto-PGF$_{1\alpha}$ requires the presence of Ca (18,19). It has been proposed that activation of phospholipase A$_2$ (PLA$_2$) may be under the control of agonist-induced increases in cellular Ca activity ([Ca]$_i$), and possibly related activation of protein kinase C (PKC) (20). The results of the few studies on agonist-stimulated PGI$_2$ release from hypertensive arteries were inconclusive. For instance, NE-stimulated release of 6-keto-PGF$_{1\alpha}$ from mesenteric arteries was either elevated or not changed in SHR depending on the control (16). A consistent increase was observed in vessels from renal hypertensive rats (16). On the other hand, NE-stimulated release of 6-keto-PGF$_{1\alpha}$ from arteries of genetically hypertensive rats was half that of the controls (21). If [Ca]$_i$ provides the primary control for arachidonic acid release, it would be expected that parallel shifts in AHR should occur in the concentration-response curves for NE-stimulation of: PLC derived second messengers (IP's), Ca-dependent functional responses (contraction and Ca dependent $^{42}$K efflux), and Ca-dependent increase in PLA$_2$ activity as measured by the arachidonic acid metabolite, 6-keto-PGF$_{1\alpha}$.

An objective of this paper is to bring together information to test this expectation on aorta from AHR. Preliminary data are also presented which indicate the presence of NE-stimulated PLD activity. The presence of such a pathway could have a significant impact on our concepts of agonist regulation since PLD activity does not result in the direct formation of IP$_3$, a major controller of Ca release (22). Evidence is also provided which indicates that the $\alpha_1$-receptor supersensitivity does not extend to the production of 6-keto-PGF$_{1\alpha}$, thus PLA$_2$ activity is not controlled by Ca levels alone.

## METHODS

Brief descriptions and references are given for techniques which we have published previously, while those which have been adapted recently are presented in more detail.

### Animals, Tissues and Solutions

Aortas were taken from male Sprague-Dawley rats which had one kidney removed and had been given saline to drink or additional infusion of aldosterone (0.25-1.0 µg/hr) by an osmotic pump (Alza, Corp., Palo Alto, CA, USA) (1,8). After four weeks treatment the systolic blood pressure averaged 190 mmHg in AHR and 120 mmHg in controls. After trimming the aorta of fat and loose connective tissue, the endothelium was removed by gently stroking the intima with filter paper moistened in the dissection solution. All solutions were similar to those used previously with a $CO_2$-$HCO_3$ buffer (PSS) employed at 37° under experimental conditions (1,8). Strips were mounted on stainless steel wires for measurements of $^{42}K$ efflux or assays of phospholipase metabolites.

## Potassium Efflux

The protocol was similar to that used previously (1,7). Strips were equilibrated 3 hr in $^{42}K$ PSS (20 µCi/ml: University of Missouri Research Reactor, Columbia, MO, USA), rinsed 1-2 sec in PSS, and serially passed through a series of vigorously gassed tubes. The protocol for the sequential application of NE was similar to that used previously (1,7,8). In one series, $[K]_o$ was increased from 5 to 25 mM by addition of KCl to the PSS. The rate constant, k, was computed for each period by a computer program, and the NE-concentration response curves were computed as described previously (1,7,8). The $EC_{50}$ was determined for each tissue by linear interpolation between the log NE concentration just above and below the 50% response.

## Inositol Phosphate Assay

Details of the procedure and its validation were given (8). Briefly, measured lengths of aorta were incubated in PSS containing [$^3H$] myo-inositol (10 µCi/ml: American Radiolabelled Chemicals, St. Louis, MO, USA) for 2 hrs, rinsed, then incubated in PSS containing 10 mM Li for 10 min before addition of NE for 30 min. The reaction was stopped by freeze clamping and placing tissues in chloroform/methanol (1:2) at 1°C for 30 min followed by agitation. Chloroform then water was added and the upper aqueous phase was used to assay for inositol phosphates. Inositol phosphates were separated by anion exchange with the resin (Dowex AGI x 8, Biorad. Corp., Richmond, CA, USA) sequentially extracted with: $H_2O$; 5 mM sodium tetraborate, 60 mM ammonium formate; 0.1 M formic acid, 0.2 M ammonium formate (IP); 0.1 M formic acid, 0.4 M ammonium formate ($IP_2$); and 0.1 formic acid, 1.0 M ammonium formate ($IP_3$). Only the $IP_2$ data are presented because the $IP_2$ response to NE was parallel to that for IP, and $IP_2$ standards indicated that sufficient $IP_2$ activity (5%) was extracted in the $IP_3$ fraction to bias the $IP_3$ data significantly. The NE stimulated [$^3H$]$IP_2$ counts per length aorta (cpm/mm) were normalized in terms of the maximal response. The $EC_{50}$ for each experiment was derived by interpolation.

## Phosphatidic Acid Assay

The details have been previously described (8). Briefly, aortic strips were incubated 2 hrs in $PO_4$-free PSS containing [$^{32}P$]$PO_4$ (50 µCi/ml, essentially $PO_4$ free, ICN, Irvine, CA, USA), with Ca reduced to 1.2 mM, followed by a 30 min incubation in normal PSS ($PO_4$ equal 1.2 mM). NE was added for 10 min and the reaction was then stopped by freeze clamping followed by homogenization in chloroform/methanol/HCL (100:100:05 by volume). The residue was reextracted three times and combined. The final PA extract was spotted on silica gel G plates (Analtech Corp., Newark, DE, USA) for thin layer chromatography using a solvent system of chloroform/pyridine/formic acid (10:6:1 by volume) (8). PA was identified by standards and TNS spray (Sigma Corp, St. Louis, MO, USA), scraped and counted on a liquid scintillation counter. The NE-stimulated [$^{32}P$]PA counts were analyzed similarly to those for $IP_2$.

## Phosphatidylethanol and Phosphatidic Acid Assay

Aortic strips of measured lengths were incubated in $[^{32}P]PO_4$ for 2 hrs and rinsed as above. The tissues were either left in the wash solution or placed for 10 min in a similar solution containing ethanol (0.5% v/v, equivalent to 109 mM), then transferred for 5 min into a similar solution containing NE (30 $\mu$M) and propranolol (3 $\mu$M). After freeze clamping, the tissues were extracted similarly to those used for inositol phosphates. The chloroform extract was spotted on Whatman LK6D prescored plates, (Whatman Lab Sales Co., Hillsboro, OR, USA) with the following solvent system: ethyl acetate/iso-octane/glacial acetic acid (9:5:2) (23). The solvents were thoroughly mixed with $H_2O$ in a separatory funnel and 100 ml upper phase (non-aqueous) was mixed with 1 ml acetic acid and used for the chamber solvent. The spots were visualized with TNS spray and by autoradiography. The bands were scraped and counted in a liquid scintillation counter and data expressed as above (cpm/mm).

## Total $[^{32}P]$Phospholipid Assay

Strips were extracted into an acidic (HCl) chloroform/methanol mixture. The phospholipids were separated on HPTLC plates (Analtech, Corp., Newark, DE, USA) which had been dipped in potassium oxalate and ethylenediaminetetra-acetic acid solution. The solvent system was chloroform/methanol/20% methylamine (60:36:10) (24). For the purpose of this study, the total $^{32}P$ activity from all the spots was normalized per strip length (cpm/mm) and was used to convert PEth and PA activity to percent of total lipid.

## 6-keto-Prostaglandin $F_{1\alpha}$ Assay

The aortic strips were equilibrated in PSS for 1 hr, then placed in either PSS or PSS + NE for 30 min. At the end of the incubation period, vessels were removed from the wires, blotted and weighed. The incubation samples were frozen at -80°C for radioimmunoassay of 6-keto-PGF$_{1\alpha}$. Preliminary studies showed that the production of 6-keto-PGF$_{1\alpha}$ was linear over this period. Radioimmunoassay kits were used for the assay (New England Nuclear, Boston, MA, USA) and the values were determined from a standard curve with corrections for the volume in the assay tube. The NE-stimulated production of 6-keto-PGF$_{1\alpha}$ was determined by subtracting the quantity formed in the absence of NE from that in its presence. Production was represented as pmol/mg wet weight produced over the 30 min period. The EC$_{50}$ was determined by a computer fit to a polynomial equation.

## Statistics

The log values of EC$_{50}$ for NE were compared because this transformation was normally distributed (7). Arithmatic means were used for other comparisons with a p value of 0.05 (based on Student's t-test) taken to be significant.

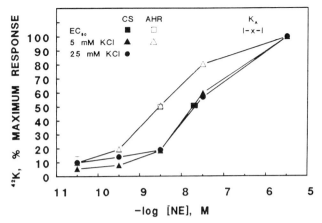

Fig. 1. Effects of NE (-log scale) on $^{42}$K efflux. Data are presented as percentage of maximal change in rate. Symbols represent means of 6 to 7 rats (Δ, AHR; ▲, control, ● control in elevated $K_o$ = 25 mM). The $EC_{50}$ is indicated by ☐ (AHR), ■ both control rats. The agonist dissociation constant, $K_A$, is indicated by x (7,8). The bars indicate one standard error of the mean (SEM) and points are joined by straight lines.

## RESULTS AND DISCUSSION

### $^{42}$K Efflux Supersensitivity

A series of experiments were conducted to determine whether partial depolarization of aorta would lead to increased sensitivity of $^{42}$K efflux to NE. As shown in Fig. 1, the $EC_{50}$ was not significantly changed in controls exposed to $[K]_o$ = 25 mM, although the basal rate was increased from 0.0098 ± 0.0004 min$^{-1}$ to 0.0143 ± 0.0007 (p < 0.001). These findings are consistent with our previous results that supersensitivity to NE was retained in AHR in the presence of Ca-channel blockers which reduced the basal $^{42}$K efflux and spontaneous contractile tension to near control levels (25,26). These results indicate that increased Ca entry which is sufficient to stimulate contraction and Ca-dependent $^{42}$K efflux does not alter NE sensitivity sufficiently to account for the decreased $EC_{50}$ in AHR. Therefore, the increased sensitivity represents a separate site which is reset in AHR, and is not just secondary to increased Ca entry by altered Ca-channel mechanisms (26).

### Phospholipase C Activity

The NE-stimulated responses of [$^3$H]IP$_2$ and [$^{32}$P]PA exhibited a significant leftward shift in AHR (Fig. 2, p < 0.001). The shift in $EC_{50}$ was similar for both metabolites (6-fold) and was similar to that observed for $^{42}$K efflux (Fig. 1). In contrast to the functional responses, the $EC_{50}$ for PLC activity were of the same order as the $K_A$ for NE. This indicated that the relation between receptor occupancy and biochemical responses approached one to one (8). The functional responses, however, were 50%

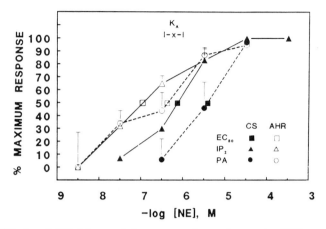

Fig. 2. Effects of NE (-log scale) on percentage change in [3H] myo-inositol bisphosphate ($IP_2$), and on [32P] phosphatidic acid ($^{32}PA$) activity in aorta from control (▲, ●) and AHR (∆, ○). The $EC_{50}$ is indicated by ■ for controls and ▢ for AHR. Six-eight rats were used for each $IP_2$ analyses while 6-16 were used for PA. Vertical bars indicate one SEM. The $K_A$ and lines are plotted as in Fig. 1. (These data are replotted from reference 8 by permission of the American Heart Association, Inc.)

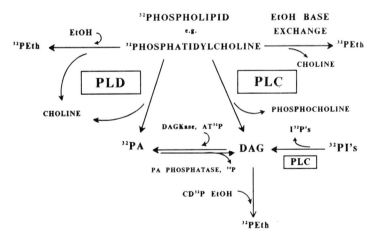

Fig. 3. Diagram showing the major pathways for phospholipase D (PLD). See text and reference 12 for details.

maximal with only 6% receptor occupancy in controls and 2% in AHR (7,8). Thus, the amplification in the signalling process occurred after the production of PLC products. For instance, relatively low levels of $IP_3$, DAG or PA would have relatively large effects on $[Ca]_i$ and resultant stimulation of $^{42}K$ efflux. Since the shifts in $EC_{50}$ for PLC activity were parallel to those for functional responses, it is reasonable to propose that supersensitivity to NE can be attributed mainly to shifts in the coupling between receptor occupancy and PLC activity. Unfortunately, because of the large variances associated with the biochemical measures of native tissues, it was not possible to demonstrate significant changes in PLC metabolites at [NE] in the $EC_{50}$ range of the functional responses. It has also been difficult to demonstrate a significant increase in $IP_3$ concentration during NE stimulation which would be consistent with the proposed action of $IP_3$ in the early release of stored Ca (Jones, unpublished observations) (27). Coburn and co-workers did, however, observe a parallel increase in the incorporation of $[^3H]$ myo-inositol into $PIP_2$ and $IP_3$ pools of rabbit aorta during a 30 min exposure to NE (27). Although some uncertainties exist in relating low concentrations and short-term effects of NE to PLC activity, it appears that supersensitivity exhibited during sustained exposure to NE has a basis in altered regulation of PLC.

Angiotensin II has been reported to cause a sustained increase in DAG in native arteries and cultured vascular smooth muscle cells (28,29), which is consistent with the role of DAG acting as a precursor for PA (Fig. 7). NE, however, did not increase total DAG under conditions in which PA levels consistently increased (29). In a preliminary study, we observed that the ratio of $[^{32}P]PA$ to inorganic P (Pi) in PA during NE stimulation was about 10 times less than the ratio of $^{32}P$ to $P_i$ in PI, PIP and $PIP_2$ (30). The $^{32}P$ to $P_i$ ratio in PA, however, was similar to that in other phospholipids such as phosphatidylcholine. The question was raised whether PLC provides a unique pathway for NE signalling in native arteries.

**Phospholipase D Activity**

Although PLD activity was demonstrated in plant cells more than 40-years ago, studies on mammalian systems are relatively recent (31). To our knowledge PLD activity has not been identified in VSM. A scheme showing the major pathways is given in Fig. 3, which is based on current information from mammalian cells (12,14). A major feature of PLD is that it catalyzes two reactions: (1) PLD acts on the terminal phosphodiester bond of glycerophospholipids to form PA and related base, e.g. choline in the case of phosphatidylcholine; and (2) PLD also promotes a transphosphatidylation reaction which can cause incorporation of a substrate such as ethanol (EtOH) onto the phosphatidyl moiety of the phospholipid such as phosphatidylcholine (Fig. 3). Base exchange can also occur spontaneously as shown in Fig. 3, however, where studied, this reaction is slow (12). An important consequence of PLD activity is that PA can be formed without first forming DAG, and from substrates such as phosphatidylcholine which are much more abundant in cell membranes than $PIP_2$.

A significant property of PLD is its activation by increased $[Ca]_i$, PKC activity as well as via receptor occupancy (14,32,33). A consensus has not been reached whether increased $[Ca]_i$ and/or PKC are prerequisites for receptor-induced activation of PLD. An important linkage between PLD and PLC pathways results from the activity of phosphatidate phosphatase to form DAG from PA (Fig. 3) which has been demonstrated in neutrophils

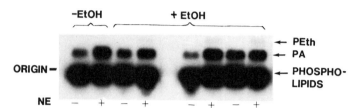

Fig. 4. Autoradiograph (1 hr exposure) of thin layer chromatography separation of phosphatidylethanol (PEth), phosphatidic acid (PA) and phospholipids. Lipids were extracted from rat aorta incubated either in the absence or presence of ethanol (0.5% v/v). Preparations stimulated by NE (30 μM) for 5 minutes are indicated by +. Although very light, identifiable bands comigrated with a PEth standard ahead of PA and other phospholipids.

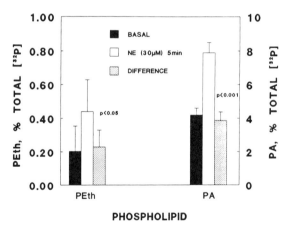

Fig. 5. Levels of [32]P labeled PEth and PA normalized as percent of total [32]P phospholipid in rat aorta (12 rats). Vertical bars represent one SEM. Note the different scales used for PEth and PA.

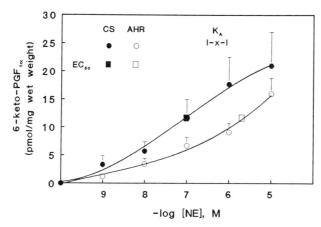

Fig. 6. Effects of NE (-log scale) on production of 6-keto-PGF$_{1\alpha}$ from rat aorta plotted as pmol/mg wet weight. Symbols represent means of 6 experiments and vertical bars are one SEM. Values represent production over a 30 minute period of exposure to NE with the basal production by control strips being subtracted. The agonist dissociation constant KA is indicated by x with + SEM shown. Curves represent computer based fits of a polynomial equation to the data.

and endothelial cells (13,15). The reverse reaction, formation of PA from DAG, has been generally accepted to exist in VSM (8,28). If PLD activity were present in VSM, then increases in DAG and PA cannot be assumed to result from PLC activity alone.

We assayed NE-stimulated PLD activity in rat aorta by measuring the product of the transphosphatidylation reaction between EtOH and phospholipids to form a novel compound, phosphatidylethanol (PEth). An autoradiograph showing the separation is given in Fig. 4. Although the [32P]PEth band was light, it was identifiable and ran clearly above that for PA and other phospholipids. NE significantly increased both [32P]PEth and [32P]PA levels in rat aorta as shown in Fig. 5. Despite the low percent of [32P]PEth in the total phospholipid pool, the changes were consistent. [32P]PA was 20-fold more abundant than [32P]PEth and the NE-stimulated changes were similar to those observed in the absence of ethanol (8). A more detailed study will be required to test the prediction that formation of PEth would compete (reduce) with the formation of PA via the PLD pathway. Based on our observation of NE stimulated PEth production, it is concluded that vascular smooth muscle contains PLD activity. Further experiments are required to determine whether direct coupling with receptors occurs or whether PLD activity is secondary to increases in [Ca]$_i$ and/or PKC activity. It is unknown whether regulation of PLD is associated with α$_1$-receptor supersensitivity in AHR. If PLD activity were linked to α$_1$-receptor occupancy via increased [Ca]$_i$ and/or PKC activity only, then parallel shifts in PLD and PLC activity and functional responses would be expected. If it is independently controlled, the shift could be in any direction.

## Phospholipase A$_2$ Activity

PLA$_2$ has been reported to be secondarily regulated by agonist-induced increases in [Ca]$_i$ and/or PKC activity (18-20). Based on this concept and the results in Figs. 1 and 2, we expected the NE-stimulated 6-keto-$PGF_{1\alpha}$ to exhibit a leftward shift in aorta from AHR. As shown in Fig. 6, however, this prediction was incorrect. The concentration-response curve was shifted to the right with the $EC_{50}$ increased about 20-fold in AHR ($p < 0.01$). The maximal production of 6-keto-$PGF_{1\alpha}$ was not significantly different between the two groups, however. While the $EC_{50}$ for AHR was similar to the $K_A$ for NE, the $EC_{50}$ for controls was significantly lower ($p < 0.05$).

The rightward shift exhibited by AHR indicates that $PLA_2$ activity and subsequent formation of cyclooxygenase (CyOx) products may be regulated by a component which is independent of NE stimulated PLC activity and subsequent increase in $[Ca]_i$ and/or PKC activity. If $PLA_2$ were regulated only by changes in $[Ca]_i$ and/or PKC activity, the NE stimulation should have been shifted to the left as was the case with other Ca-dependent functions, e.g. $^{42}K$ efflux in Fig. 1. Since the maximal NE-stimulated production was similar in both groups, it is not apparent that major attenuation of $PLA_2$ activity occurred. Furthermore, $PGI_2$ synthase activity and basal release of 6-keto-$PGF_{1\alpha}$ was increased from aorta from DOCA-salt hypertensive rats (34). Therefore, limited production of $PGI_2$ from its precursor is not well supported in AHR. On the other hand, $PLA_2$ and PLC activity assayed on membrane homogenates was reduced (35 and 20%, respectively) in aorta from the same animals (34). Receptor stimulated responses, however, were not reported in this study. Based on our results (Fig. 2, Ref. 8) the NE- stimulated production of radio-labeled phospholipid products (IP, $IP_2$, PA) was not impaired in AHR (a similar model to DOCA). Therefore, it is not clear that a direct relation exists between the assay of maximal PLC (as well as $PLA_2$) activity in membrane fragments and agonist-induced changes in intact cells. Although the rightward shift in NE stimulated 6-keto-$PGF_{1\alpha}$ could result from reduced $PLA_2$ activity, it could also represent a decreased sensitivity of $PLA_2$ to $[Ca]_i$ and/or PKC activity or even a shift in the relative activity of the intermediary enzymes associated with the cyclooxygenase pathway. Although speculative, $PLA_2$ may also be directly controlled by $\alpha_1$-receptor occupancy (Fig. 7). Uncoupling of direct receptor control in AHR could underlie the rightward shift in NE-stimulated 6-keto-$PGF_{1\alpha}$ which would then be controlled mainly by increased $[Ca]_i$ and/or PKC activity in response to $\alpha_1$-receptor stimulation of PLC.

**Summary**

Many of the concepts presented in this paper are summarized in Fig. 7. Some aspects are well supported while others are speculative (marked by ?). The operation of PLC in VSM is well established, and in some hypertensive models (AHR, SHRSP) PLC assays exhibited altered activation. Currently this pathway leading to the production of $IP_3$ and DAG is considered to be the major regulator of Ca release from sarcoplasmic reticulum (SR) and Ca entry by channels (CaC). Regulation of PKC by $[Ca]_i$ and DAG is thought to play a major role in controlling Ca entry. PKC has also been proposed to regulate $PLA_2$ as well as PLD in conjunction with elevated $[Ca]_i$. An important issue to be resolved is whether receptor regulation of other lipases occurs independently of the PLC-$[Ca]_i$-PKC axis. Currently information supporting receptor regulation is lacking for VSM, but few studies have been conducted.

Our observation that NE stimulation of PLD activity occurs in VSM indicates that the control of VSM by biochemical messengers is much more complicated than previously proposed. This seemingly redundant pathway may allow VSM to use alternate substrates for producing PA and DAG than are readily available to PLC. It also allows PA to be produced directly without phosphorylation of DAG. Although the role of PA in the regulation of Ca entry was proposed earlier, definitive studies establishing this linkage are still required. Any PLD activity on $PIP_2$ would produce biochemical messengers (PA, DAG) which could stimulate Ca entry without producing the messenger, $IP_3$, associated with Ca release (inactive $IP_2$ would be produced, Fig. 7). If PLC and PLD were independently regulated by receptor-guanine nucleotide-regulatory protein (G-protein) complexes, this would offer the potential for some agonists to excite VSM by Ca release and Ca entry mechanisms while others may excite by Ca entry alone. This system would also circumvent the problem of limited substrate for cellular regulation of $[Ca]_i$ if $PIP_2$ were the primary substrate. This limitation does not exist with other phospholipids such as phosphatidylcholine which is a preferred substrate for PLD. The presence of multiple phospholipases under separate receptor regulation allows for a wider range of tissue responses to various agonists, than a system which is linked only through the PLC-$[Ca]_i$-PKC axis.

The presence of a PLD pathway also reopens the interpretation of previous studies which demonstrated a resetting between receptor occupancy and production of second messengers by PLC (Fig. 7). Since a clear shift in the concentration-response curves has not been demonstrated for $IP_3$, the reliance on other metabolic products (although consistent with the hypothesis) is not definitive. As shown in Fig. 3, significant crossover exists between PLD and PLC production of PA. The potential is shown in Fig. 7 for $IP_2$ (and IP) to be formed by the action of PLD on $PIP_2$ (and PIP). Although resetting of the receptor-PLC coupling process remains a likely candidate for $\alpha_1$-adrenergic receptor-supersensitivity, additional consideration should be placed on the relative role of PLD in the coupling process and whether it is reset in AHR.

Altered Ca entry is a second established abnormality in AHR which could indirectly contribute to NE supersensitivity by raising basal $[Ca]_i$ (Fig. 7). In addition to increasing functional activity (Ca-dependent K efflux, contraction) elevated $[Ca]_i$ could enhance the activity of PLD, $PLA_2$ and PKC to modulate the production of biochemical messengers (Fig. 7). Selectively increasing Ca entry into aorta from control rats, however, did not increase the sensitivity to NE (Fig. 1). Although increased basal entry of Ca in AHR has important effects on basal contractile and ion flux levels, it does not appear to have a major effect on $\alpha$-adrenergic receptor supersensitivity.

The rightward shift of the NE-stimulated production of 6-keto-$PGF_{1\alpha}$ shows that the sensitivity changes in AHR are not stereotypical. This finding supports the concept that $PLA_2$ may be controlled by mechanism(s) in addition to PLC-$[Ca]_i$-PKC axis (Fig. 7). The reduced production of $PGI_2$ (a relaxant of NE stimulated contraction) may enhance NE responses and contribute to increased vascular resistance in AHR. Further study is required, however, to determine the relative importance of arachidonic acid products released by VSM as modulators of agonist-induced contraction. A need also exists to explore whether shifts occur in lipoxygenase and endoperoxidase products.

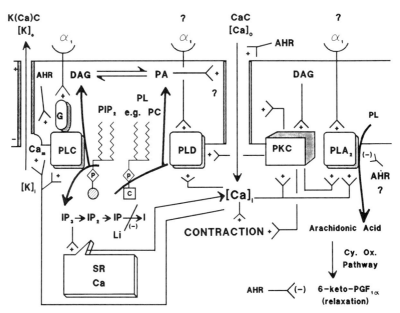

Fig. 7. Schematic model for $\alpha_1$-adrenergic receptor ($\alpha_1$) regulation of vascular smooth muscle phospholipase activity. See text for detailed explanation. Abbreviations: K (Ca) C, Ca-dependent K channels; CaC, calcium channels; AHR, aldosterone-salt hypertensive rat; PLC, PLD, PLA$_2$, phospholipases C,D and A$_2$ respectively; PKC, protein kinase C;G, G-protein; DAG, diacylglycerol; PA, phosphatidic acid; PIP$_2$, phosphatidylinositol-4,5-bisphosphate; I, inositol; IP, inositol phosphate; IP$_2$, inositol bisphosphate; IP$_3$, inositol trisphosphate; PL, phospholipid; PC, phosphatidylcholine; C, choline; Li, lithium; Cy Ox, cyclooxygenase; SR, sarcoplasmic reticulum; [Ca]$_i$, free cytosolic calcium activity; Ca$_m$, Ca activity at intracellular membrane.

Although supersensitivity to agonists has been associated with various forms of hypertension, only recently has the complexity of the problem been appreciated. We are still in a reiterative process of establishing biochemical mechanisms and determining their involvement in the pathophysiology of hypertension. Hopefully, as we work through this process in VSM two important issues will be resolved; the relative importance of PLD and $PLA_2$ in biochemical signalling and whether one or both can be controlled directly by receptor mechanisms.

## ACKNOWLEDGEMENTS

The authors thank Nancy N. Cook for expert assistance. This work was supported by US Public Health Service grants HL-30519 and HL-15852. S.B. Jones is supported by training grant HL-07094.

## REFERENCES

1. Garwitz ET, Jones AW. Aldosterone infusion into the rat and dose-dependent changes in blood pressure and arterial ionic transport. *Hypertension* 4: 374, 1982.
2. Holloway ET, Bohr DF. Reactivity of vascular smooth muscle in hypertensive rats. *Circ Res* 33: 678, 1973.
3. Jones AW. Altered ion transport in vascular smooth muscle from spontaneously hypertensive rats: influence of aldosterone, norepinephrine, and angiotensin. *Circ Res* 33: 563, 1973.
4. Berecek KH, Strockes M, Gross F, Changes in renal vascular sensitivity at various stages of deoxycorticosterone hypertension in rats. *Circ Res* 46: 619, 1980.
5. Katovich MJ, Soltis EE, Iloye E, Field FP. Time course alteration in vascular adrenergic responsiveness in DOCA/NaCl treated rat. *Pharmacology* 29: 173, 1984.
6. Jones SB, Smith JM, Jones AW, Bylund DB. Alpha-1 adrenergic receptor binding in aorta from rat and dog: comparison of [$^3$H] prazosin and ß-iodo-[$^{125}$I]-4-hydroxyphenyl-ethyl-amino-methyl-tetralone. *J Pharmacol Exp Ther* 241: 875, 1987.
7. Smith JM, Jones SB, Bylund DB, Jones AW. Characterization of the alpha-1 adrenergic receptor in the thoracic aorta of control and aldosterone hypertensive rats: correlation of radio-ligand binding with potassium efflux and contraction. *J Pharmacol Exp Ther* 241: 882, 1987.
8. Jones AW, Geisbuhler BB, Shukla SD, Smith JM. Altered biochemical and functional responses in aorta from hypertensive rats. *Hypertension* 11: 627, 1988.
9. Exton JH. Mechanisms of action of calcium-mobilizing agonists: some variations on a young theme. *FASEB J* 2: 2670, 1988.
10. Turla MB, Webb RC. Augmented phosphoinositide metabolism in aortas from genetically hypertensive rats. *Am J Physiol* 258: H173, 1990.
11. McMahon EG, Paul RJ. Calcium sensitivity of isometric force in intact and chemically skinned aortas during the development of aldosterone-salt hypertension in the rat. *Circ Res* 56: 427, 1985.
12. Pai JK, Siegel ME, Egan RW, Billah MM. Phospholipase D catalyzes phospholipid metabolism in chemotactic peptide-stimulated HL-60 granulocytes. *J Biol Chem* 263: 12472, 1988.

13. Billah MM, S. Eckel, Mullmann TJ, Egen RW, Siegel MI. Phosphatidylcholine hydrolysis by phospholipase D determines phosphatidate and diglyceride levels in chemotactic peptide-stimulated human neutrophils. *J Biol Chem* 264: 17069, 1989.

14. Exton JH. Signaling through phosphatidylcholine breakdown. *J Biol Chem* 205: 1, 1990.

15. Martin TW. Formation of diacylglycerol by a phospholipase D - phosphatidate phosphatase pathway specific for phosphatidylcholine in endothelial cells. *Biochim Biophys Acta* 962: 282, 1988.

16. Desjardins-Giasson S, Gutkowska J, Garcia B, Genest J. Release of prostaglandins by the mesenteric artery of the renovascular and spontaneously hypertensive rat. *Can J Physiol Pharmacol* 62: 89, 1984.

17. Nishimiya T, Daniell HB, Webb JB, Oatis J, Walle T, Gaffney TE, Halushka PV. Chronic treatment with propranolol enhances the synthesis of protaglandins $E_2$ and $I_2$ by the aorta of spontaneously hypertensive rats. *J Pharmacol Exp Ther* 253: 207, 1990.

18. Jeremy JY, Mikhailidis DP, Dandona P. Adrenergic modulation of vascular prostacyclin ($PGI_2$) secretion. *Eur J Pharmacol* 114: 33, 1985.

19. Stewart D, Pountney E, Filchett D. Norepinephrine-stimulated vascular prostacyclin synthesis: Receptor-dependent calcium channels control prostaglandin synthesis. *Can J Physiol Pharmacol* 62: 1341, 1984.

20. Rana RS, Hokin LE. Role of phosphoinositides in transmembrane signaling. *Physiol Rev* 70: 115, 1990.

21. Pipili E, Poyses NL. Release of prostaglandins $I_2$ and $E_2$ from the perfused mesenteric arterial bed of normotensive and hypertensive rats. Effects of sympathetic nerve stimulation and norepinephrine administration. *Prostaglandins* 23: 543. 1982.

22. Somlyo AV, Kitazawa T, Horiuti K, Kobayashi S, Trentham D, Somlyo AP. Heparin-sensitive inositol triphosphate signaling and the role of G-protein in $Ca^{2+}$ release and contractile regulation in smooth muscle. In: *Frontiers in Smooth Muscle Research*. New York: Alan R. Liss, 1990, pp 167.

23. Lapetina EG, Siess W. Measurement of inositol phospholipid turnover in platelets. *Methods in Enzymology* 141: 176, 1987.

24. Shukla SD, Hanahan DJ. AGEPC (platetet activating factor) induced stimulation of rabbit platelets: Effects on phosphatidyl-inositol, di- and tri- phosphoinositides and phosphatidic acid metabolism. *Biochem Biophys Res Commun* 106: 697, 1982.

25. Jones AW, Smith JM. Altered Ca-dependent fluxes of [42]K in rat aorta during aldosterone-salt hypertension. *Prog Clin Biol Res* 219: 265, 1986.

26. Smith JM, Jones AW. Calcium antagonists inhibit elevated potassium efflux from aorta of aldosterone-salt hypertensive rats. *Hypertension* 15: 78, 1990.

27. Coburn RF, Baron C, Papadoulos MT. Phosphoinositide metabolism and metabolism - contraction coupling in rabbit aorta. *Am J Physiol* 299: H1476, 1988.

28. Griendling KK, Rittenhouse SE, Brock TA, Ekstein LS, Gimbrone MA, Jr., Alexander RW. Sustained diacylglycerol formation from inositol phospholipids in angiotensin II-stimulated vascular smooth muscle cells. *J Biol Chem* 261: 5901, 1986.

29. Ohanian J, Ollerenshaw J, Collins P, Heagerty A. Agonist-induced production of 1,2-diacylglycerol and phosphatidic acid in intact resistance arteries. *J Biol Chem* 265: 8921, 1990.

30.  Jones AW, Shukla SD, Geisbuhler BB. Evidence for phospholipase D activity in rat aorta that is stimulated by norepinephrine. *FASEB J* 4: A333 Abs., 1990.
31.  Heller M. Phospholipase D. *Adv Lipid Res* 16: 267, 1978.
32.  Reinhold SL, Prescott SM, Zimmerman GA, McIntyre TM. Activation of human neutrophil phospholipase D by three separable mechanisms. *FASEB J* 4: 208, 1990.
33.  Martin TM, Feldman DR, Goldstein KE, Wagner JR. Long-term phorbol ester treatment dissociates phospholipase D activation from phosphoinositide hydrolysis and prostacyclin synthesis in endothelial cells stimulated with bradykinin. *Biochem Biophys Res Commun* 165: 319, 1989.
34.  Uehara Y, Ishimitsu T, Ishii M, Sugimoto T. Prostacyclin synthase and phospholipases in the vascular wall of experimental hypertensive rats. *Prostaglandins* 34: 423, 1987.

# ROLE OF CONTRACTILE AGONISTS IN GROWTH REGULATION
## OF VASCULAR SMOOTH MUSCLE CELLS

Gary K. Owens

Department of Physiology
University of Virginia School of Medicine
Charlottesville, VA 22908

## INTRODUCTION

Accelerated smooth muscle growth is a characteristic feature in arteries of hypertensive patients and animals (1,15,26,27,36,40). In resistance vessels, this accelerated growth of smooth muscle contributes to development of medial thickening or hypertrophy and is thought to play an important role in the etiology of hypertension (1,13). It is hypothesized that medial hypertrophy confers a mechanical advantage such that at any given level of smooth muscle activation vascular resistance is greater in hypertensives than in controls (12,25). It should be noted, however, that medial hypertrophy does not appear to occur in very small arteries (< 100 μm internal diameter) and arterioles, but rather appears confined to large and intermediate sized resistance arteries (i.e. 150-300 μm internal diameter) (8,19,26,40). Extensive medial hypertrophy occurs in large conduit vessels, such as the aorta (33). In these vessels, medial hypertrophy is thought to be an adaptive process to minimize changes in wall stress that occur as a consequence of increases in blood pressure (34).

The cellular nature of the smooth muscle cell growth response, at least in the spontaneously hypertensive rat (SHR), has been shown to be quite different in intermediate size resistance vessels versus large vessels. In the SHR, medial thickening in the aorta is due almost entirely to an increase in the size of pre-existing smooth muscle cells, or cellular hypertrophy, rather than cellular proliferation or hyperplasia (33,38,39). Interestingly, smooth muscle cell hypertrophy is accompanied by DNA endoreduplication and development of polyploidy in a large fraction of the smooth muscle cells in the vessel wall (33). In contrast, medial thickening in intermediate size mesenteric resistance vessels in the SHR is due to smooth muscle cell hyperplasia (26,40).

The factors which mediate smooth muscle growth responses in vessels of hypertensive animals have not been clearly identified. This chapter will focus on consideration of the possible role of contractile agonists (e.g. angiotensin II, norepinephrine, and arginine vasopressin) in mediation of growth responses of smooth muscle cells in hypertension, as well as in developmental growth, differentiation, and maintenance of vascular mass in normal animals.

## EVIDENCE IMPLICATING A ROLE FOR CONTRACTILE AGONISTS IN MEDIATION OF SMOOTH MUSCLE CELL GROWTH IN HYPERTENSION

There is considerable circumstantial evidence implicating a role for increased blood pressure, wall stress, or other mechanical factor in the growth response of smooth muscle cells in hypertension. We (34,35) and others (24) have consistently observed a high degree of correlation between the level of blood pressure elevation and the extent of vascular hypertrophy. However, there is no definitive evidence for a causal role for these factors in the growth response and one cannot rule out the possibility that the actual growth mediator might be a covariant of blood pressure, for example, those factors which mediate vascular tone (i.e. contractile agonists) and thereby control blood pressure. Indeed there is convincing evidence that tissue and/or circulating levels of contractile agonists are elevated in many models of hypertension (2,46). Furthermore, drug intervention studies in the SHR have implicated a possible role for angiotensin II in mediation of smooth muscle hypertrophy (14,35). We compared the effects of a variety of antihypertensive drugs on development of smooth muscle cell hypertrophy and hyperploidy in the SHR. Results demonstrated that captopril (a converting enzyme inhibitor) was more effective in preventing hypertrophy than was hydralazine (a direct smooth muscle cell relaxant) for a given level of blood pressure lowering. Results suggest that hypertrophy is not simply a response to increased blood pressure and implicate a role for angiotensin II in the growth response. Interestingly, captopril was also effective in reducing aortic smooth muscle content in Wistar-Kyoto rats, suggesting that angiotensin II may also play some role in the maintenance of vascular mass in normotensive rats.

The idea that contractile agonists may have growth modulatory properties *in vivo* is intriguing, since it may provide a mechanism whereby smooth muscle tissues might modulate their mass in accordance with their work load. Whereas at present there is no direct evidence showing that such is the case, there is convincing evidence that contractile agonists have growth promoting properties in cultured smooth muscle cells (5,17,18).


## ANGIOTENSIN II AND ARGININE VASOPRESSIN INDUCE HYPERTROPHY OF CULTURED SMOOTH MUSCLE CELLS

We have previously demonstrated that angiotensin II and arginine vasopressin are potent hypertrophic agents for cultured rat aortic smooth muscle cells eliciting a 20-40% increase in protein content and cell size following chronic treatment for four days in a defined serum-free medium (17,18). Increases in protein content were concentration dependent, were inhibited with specific receptor antagonists, and in the case of angiotensin II were associated with increases in the fraction of polyploid cells as occurs in hypertrophic smooth muscle cells *in vivo*. We saw no evidence that agonists were mitogenic, either when administered alone, or in combination with serum or platelet-derived growth factor (PDGF). Agonist treatment induced a rapid and sustained increase in [35]S-methionine incorporation into TCA precipitable material, but no change in fractional release of [14]C-tyrosine from prelabeled cells, indicating that cellular hypertrophy was due to an increase in protein synthesis rather than a change in the overall rate of protein degradation (17). Our results with angiotensin II have been confirmed and extended by a number of other groups (5,44).

## MECHANISMS WHEREBY CONTRACTILE AGONISTS STIMULATE PROTEIN SYNTHESIS

Relatively little is known regarding the mechanisms whereby contractile agonists stimulate protein synthesis and hypertrophy of smooth muscle cells. One experimental approach is to focus on examination of the role of the various early signal transduction pathways and messengers, such as $Ca^{+2}$, $Na^+/H^+$ exchange, inositol phosphates, diacylglycerol, protein kinase C, etc. in mediation of the hypertrophic response. This approach has been employed by a number of investigators (5,22,28,44,45) and has yielded useful information regarding the mechanisms for angiotensin II-induced hypertrophy (reviewed in 33). Major findings of these studies are summarized below.

1.  Angiotensin II-induced increases in protein synthesis can be blocked with the transcriptional inhibitors alpha-amanitin or actinomycin D (Owens and Geisterfer, unpublished data, 5). Although these data suggest that angiotensin II and arginine vasopressin-induced increases in protein synthesis are dependent on transcription, they provide no definitive insight as to what genes are involved, and are subject to criticisms regarding non-specific actions of these drugs, as well as drug-induced depletion of various proteins, ribosomal subunits, etc. that comprise the protein synthetic machinery of the cell.

2.  Angiotensin II-induced increases in protein synthesis are dependent on agonist-induced changes in cytosolic $Ca^{+2}$, but cannot be reproduced simply by elevating intracellular $[Ca^{+2}]$ with a $Ca^{+2}$ ionophore (5). It is thus unclear whether the role of $Ca^{+2}$ is permissive or regulatory.

3.  Angiotensin II-induced increases in protein synthesis are unaffected by treatment of cells with dimethylamiloride (5) indicating that they are not dependent on $Na^+/H^+$ exchange as suggested by Scott-Burden *et al* (44).

4.  Phorbol ester induced down-regulation of protein kinase C had no effect on angiotensin II induced increases in protein synthesis suggesting that changes are not dependent on this pathway (5).

5.  Angiotensin II has been shown to stimulate marked increases in expression of the protooncogenes *c-fos* (22,29,45) and *c-myc* (28), raising the possibility that the effects of angiotensin II on growth are mediated by the protein products of these genes. However, while these are clearly growth related genes, the function of their protein products in normal cells has not been clearly defined (20), and there is currently no direct evidence supporting a role for these protooncogenes in the hypertrophic response.

6.      Dzau and co-workers (27,28) have demonstrated that angiotensin II treatment of cultured rat aortic smooth muscle cells increases expression of PDGF A-chain mRNA and PDGF secretion into the media. This increase is dependent on protein synthesis and occurs approximately six hours following angiotensin II stimulation. Since no B-chain PDGF mRNA was detectable, these observations raise the interesting possibility that angiotensin II-induced increases in protein synthesis and subsequent cellular hypertrophy may be mediated via stimulation of PDGF AA production. However, it remains to be determined whether PDGF AA induces hypertrophy rather than hyperplasia of cultured rat aortic smooth muscle cells.

A major limitation in investigations of the role of various second messenger systems in the hypertrophic response is that it is extremely difficult, if not impossible, to show a direct causal relationship (or in some cases any relationship at all) between the early change in the second messenger system and a relatively long-term change in protein metabolism. One difficulty is that many of the experimental manipulations of second messenger systems, when carried out for long periods of time, are likely to have direct effects on protein synthesis (e.g. depletion of intracellular $Ca^{+2}$), thereby confounding data interpretation. Finally, an important consideration when using this experimental approach is that one must demonstrate that the proteins which contribute to the early increases in protein synthesis actually contribute to the cellular hypertrophy observed in chronic experiments, since it is conceivable that many of the proteins whose synthesis is elevated in response to angiotensin II may have nothing to do with the long-term hypertrophic response.

An alternative approach is to identify those proteins which accumulate in hypertrophic cells, and to determine the mechanisms responsible for their accumulation. Note that the emphasis here is not initially on identification of important regulatory proteins but rather on identification of the major proteins (presumably structural) which account for the contractile agonist-induced cellular hypertrophy. One can then select a representative group of these proteins and pursue the molecular mechanisms that contribute to their increased expression. Determining whether increases in protein content are general in nature or restricted to a selected subset of proteins may provide insight into potential mechanisms responsible for the hypertrophy, as well as possible functional alterations in hypertrophic cells.

We have recently completed studies in which we utilized a combination of one-dimensional and two-dimensional gel electrophoretic analyses to determine what proteins are increased in rat aortic smooth muscle cells treated chronically with angiotensin II or arginine vasopressin (47). Results demonstrated the following:

1.      Synthesis of most, if not all, proteins was increased, suggesting that at least part of the effect of agonists on protein synthesis involved a generalized mechanism for increasing protein synthesis such as an increase in synthesis of ribosomal RNA or proteins, increased elongation rates, etc.

2.      Increases in synthesis and content of certain cytoskeletal (e.g. vimentin) and contractile proteins (e.g. actin, alpha-tropomyosin), however, far-exceeded overall increases in protein synthesis and content indicating that agonists also had selective effects on protein expression.

3.  Increases in actin synthesis and content were accompanied by increases in actin mRNA content indicating that changes are not mediated solely by translational control mechanisms.

4.  Angiotensin II and arginine vasopressin stimulated selective increases in the smooth muscle specific contractile protein, smooth muscle alpha-actin, suggesting that contractile agonists may also act as important positive regulators of smooth muscle differentiation. This is in marked contrast to the effects of the potent smooth muscle mitogen, PDGF, which induces a rapid and dramatic decrease in smooth muscle alpha-actin expression (7,11).

These results support a possible role for angiotensin II and arginine vasopressin in mediation of smooth muscle cell hypertrophy. Furthermore, results suggest that contractile agonists may also play an important role in regulating developmental growth and differentiation of smooth muscle as well as in mediation of normal adaptive growth responses of smooth muscle tissues as a means of matching the contractile mass of the tissue to its functional demands. Additional studies will be required to determine whether mRNAs for contractile proteins other than actin are also increased, and whether changes are mediated transcriptionally or post-transcriptionally.

## ROLE OF CONTRACTILE AGONISTS AS MITOGENS

As discussed earlier, medial hypertrophy in intermediate-sized mesenteric resistance vessels of the SHR is due to proliferation of smooth muscle cells or hyperplasia rather than cellular hypertrophy. Furthermore, there is clear epidemiological evidence demonstrating that hypertension is a major risk factor for development of atherosclerosis (21), a disease characterized by intimal proliferation of smooth muscle cells (43). Given that circulating or tissue levels of a variety of contractile agonists are elevated in many models of hypertension (2,46), it is of interest to consider whether contractile agonists might also stimulate smooth muscle cell proliferation under some circumstances.

As noted earlier, a number of studies including our own, demonstrated that angiotensin II had little or no mitogenic activity for cultured rat aortic smooth muscle cells, either when administered alone (17), in combination with other growth factors (17), or with low concentrations of serum (5). Furthermore, we have observed similar results using the contractile agonists arginine vasopressin (18), and norepinephrine (unpublished observations). However, in contrast to our findings, Campbell-Boswell *et al* (9) found that angiotensin II and arginine vasopressin induced a mitogenic response in explant-derived cultured human aortic smooth muscle cells. Furthermore, a number of studies have shown proliferative responses of smooth muscle as well as non-muscle cells (principally fibroblasts) in response to a variety of contractile agonists including arginine vasopressin (42), epinephrine (6), bradykinin (32), prostaglandin $E_1$ (31), and serotonin (30). In general, the reported effects of contractile agonists observed in these studies were relatively small as compared to growth factors such as PDGF, and little or no proliferative response was observed in quiescent cells that were growth-arrested using either serum starvation or plasma derived serum, the two common methods for inducing quiescence *in vitro*. Furthermore, given the considerable overlap in second messenger systems stimulated by contractile agonists and classic growth factors such as PDGF, fibroblast growth factor, and epidermal growth factor, it is not surprising that contractile agonists may show some mitogenic activity (reviewed in 33). Note that the converse is also true in that PDGF, and a number of growth factors have been shown to induce contraction of smooth muscle although their efficacy as contractile agonists was extremely low (3,4).

It seems likely that many, if not all of the differences in results obtained in studies of the role of contractile agonists as mitogens relate to differences in experimental protocols or in the cells assayed. For example, the discrepancies between our results and those of Campbell-Boswell *et al* (9) may reflect differences between species, sera, or the differentiational status of the cultured smooth muscle cells employed. Under the conditions of our studies, rat aortic smooth muscle cells express high levels of smooth muscle specific contractile proteins including smooth muscle alpha-actin (7,11,37) and smooth muscle myosin heavy chain (41), while the explant-derived smooth muscle cells used by Campbell-Boswell *et al* (9) were most likely highly modulated and expressed few, if any, proteins characteristic of differentiated smooth muscle (10). Thus, the proliferative growth response they observed, as well as the proliferative responses observed in fibroblast systems (42), may be due to the hyperresponsive nature of these cells to growth factors as compared to ours due to loss of growth suppressor mechanisms associated with cellular differentiation. This is not to say that studies of highly modulated cells are not relevant with regard to smooth muscle growth regulation *in vivo*. Smooth muscle cells in atherosclerotic lesions are known to be highly modulated with respect to expression of cell-specific contractile proteins (16,23), and under these conditions may show increased growth responsiveness to contractile agonists. Furthermore, given the proliferative response of the microvascular smooth muscle in the SHR (26,40), one must also consider the possibility of differential growth responses of smooth muscle cells derived from different vascular beds. However, this is largely speculation, and current evidence strongly indicates that classic contractile agonists such as angiotensin II, norepinephrine, and arginine vasopressin are, at best, very poor mitogens for truly quiescent, at least partially differentiated smooth muscle cells in culture (33).

## SUMMARY AND CONCLUSIONS

There is now clear evidence demonstating that contractile agonists such as angiotensin II and arginine vasopressin are potent hypertrophic agents for cultured vascular smooth muscle cells. Furthermore, there is circumstantial evidence supporting a role for these factors in mediation of smooth muscle cell hypertrophy in hypertensive animal models as well as in maintenance of contractile mass in normotensive animals. At least part of the hypertrophic effect of angiotensin II and arginine vasopressin appears to involve a generalized increase in protein synthesis since the synthesis of most if not all proteins is increased to some extent. However, in addition, these agonists also stimulate large selective increases in the synthesis and content of a number of cytoskeletal and smooth muscle cell specific contractile proteins, including smooth muscle alpha-actin. The latter result is quite exciting since it suggests that contractile agonists may play an important role in regulation of developmental growth and differentiation of vascular smooth muscle as well as in modulating the contractile mass of smooth muscle tissues in accordance with functional demands. Observations that agonists increase expression of smooth muscle alpha-actin mRNA show that hypertrophy is not regulated solely at the translational level, although it remains to be determined whether changes are mediated transcriptionally and/or post-transcriptionally. In any event, further examination of the mechanisms whereby contractile agonists alter expression of these proteins should provide important insight regarding how these factors act as hypertrophic agents. The signal transduction pathways which are important in mediation of growth effects are not well understood, and nothing is known with regard to the role of mechanical factors in the contractile agonist-induced hypertrophic response of cultured cells. Finally, very little is known with regard to factors that mediate smooth muscle cell hypertrophy and hyperplasia *in vivo* or the interrelationships between hyperplastic and hypertrophic growth of smooth muscle cells.

## ACKNOWLEDGMENT

This work was supported by Public Health Service Grants PO1-HL19242 and RO1-HL38854 from the National Institutes of Health.

## REFERENCES

1. Aalkjaer C, Heagerty AM, Petersen KK, Swales JD, Mulvany MJ. Evidence for increased media thickness, increased neuronal amine uptake, and depressed excitation-contraction coupling in isolated resistance vessels from essential hypertensives. *Circ Res* 61: 181-6, 1987.
2. Asaad M, Antonaccio M. Vascular wall renin in spontaneously hypertensive rats - potential relevance to hypertension maintenance and antihypertensive effect of captopril. *Hypertension* 4: 487-493, 1982.
3. Berk B, Alexander R, Brock T, Gimbrone M, Webb R. Vasoconstriction: A new activity for platelet-derived growth factor. *Science* 232: 87-90, 1986.
4. Berk BC, Brock TA, Webb RC, Taubman MB, Atkinson WJ, Gimbrone MAJ, Alexander RW. Epidermal growth factor, a vascular smooth muscle mitogen, induces rat aortic contraction. *J Clin Invest* 75: 1083-1086, 1985.
5. Berk BC, Vekshtein V, Gordon HM, Tsuda T. Angiotensin II-stimulated protein synthesis in cultured vascular smooth muscle cells. *Hypertension* 13: 305-314, 1989.
6. Blaes N, Boissel J. Growth stimulating effect of catecholamines on rat aortic smooth muscle cells in culture. *J Cell Physiol* 116: 167-172, 1983.
7. Blank R, Thompson M, Owens G. Cell cycle versus density dependence of smooth muscle alpha actin expression in cultured rat aortic muscle cells. *J Cell Biol* 107: 299-306, 1988.
8. Bohlen H, Lohach D. In vivo study of microvascular wall characteristics and resting control of young and mature spontaneously hypertensive rats. Blood Vessels 15: 322-330, 1978.
9. Campbell-Boswell M, Robertson A. Effects of angiotensin II and vasopressin on human smooth muscle cells in vitro. *Expt Molec Path* 35: 265-276, 1981.
10. Chamley-Campbell JH, Campbell GR, Ross R. Phenotype-dependent response of cultured aortic smooth muscle to serum mitogens. *J Cell Biol* 89: 379-383, 1981.

11. Corjay M, Lynch K, Thompson M, Owens G. Differential effect of PDGF versus serum induced growth on SM alpha actin and beta non-muscle actin mRNA expression in cultured rat aortic smooth muscle cells. *J Biol Chem* 264: 10501-10506, 1989.

12. Folkow B. Constriction-distension relationships of resistance vessels in normo- and hypertension. *Clin Sci* 57: 23s-25s, 1979.

13. Folkow B. Physiological aspects of primary hypertension. *Physiol Rev* 62: 347-504, 1982.

14. Freslon J, Giudicelli J. Compared myocardial and vascular effects of captopril and dihydralazine during hypertension development in spontaneously hypertensive rats. *Br J Pharmac* 80: 533-543, 1983.

15. Furuyama M. Histometrical investigations of arteries in reference to arterial hypertension. *Toihoku J Exper Med* 76: 388-414, 1962.

16. Gabbiani GE, Rungger-Brandle DeChastonay C, Franke WW. Vimentin containing smooth muscle cells in aortic intimal thickening after endothelial injury. *Lab Invest* 47: 265-269, 1982.

17. Geisterfer A, Peach MJ, Owens GK. Angiotensin II induces hypertrophy, not hyperplasia of cultured rat aortic smooth muscle cells. *Circ Res* 62: 749-756, 1988.

18. Geisterfer A, Owens G. Arginine vasopressin induced hypertrophy of cultured rat aortic smooth muscle cells. *Hypertension* 14: 413-420, 1989.

19. Ichijima K. Morphological studies on the peripheral small arteries of spontaneously hypertensive rats. *Jpn Circ J* 33: 785-813, 1969.

20. Kaczmarek L. Protooncogene expression during the cell cycle. *Lab Invest* 54: 365-76, 1986.

21. Kannel W, Doyle J, Ostfield A, Jenkins C, Kuller L. Podell R, Stamler J. Optimal resources for primary prevention of atherosclerotic diseases. *Circ* 70: 157A-205A, 1984.

22. Kawahara Y, Sunako M, Tsuda T, Fukuzaki H, Fukumoto Y, Takai Y. Angiotensin II induces expression of the c-fos gene through protein kinase C activation and calcium ion mobilization in cultured vascular smooth muscle cells. *Biochem Biophys Res Commun* 150: 52-9, 1988.

23. Kocher O, Skalli O, Bloom W, Gabbiani G. Cytoskeleton of rat aortic smooth muscle cells: Normal conditions and experimental intimal thickening. *Lab Invest* 50: 645-652, 1984.

24. Lichtenstein A, Brecher P, Chobanian A. Effects of hypertension and its reversal on the size and DNA content of rat aortic smooth muscle cells. *Hypertension* 8: 1150-1154, 1986.

25. Lopez RA, Mendoza SA, Nanberg E, Sinnett SJ, Rozengurt E. $Ca^{2+}$-mobilizing actions of platelet-derived growth factor differ from those of bombesin and vasopressin in Swiss 3T3 mouse cells. *Proc Natl Acad Sci USA* 84: 5768-72, 1987.

26. Mulvany M, Baandrup U, Gundersen H. Evidence for hyperplasia in mesenteric resistance vessels of spontaneously hypertensive rats using a three-dimensional disector. *Circ Res* 57: 794-800, 1985.

27. Mulvany MJ, Hansen PK, Aalkjaer C. Direct evidence that the greater contractility of resistance vessels in spontaneously hypertensive rats is associated with a narrowed lumen, a thickened media, and an increased number of smooth muscle cell layers. *Circ Res* 43: 854-864, 1978.

28. Naftilan AJ, Pratt RE, Dzau VJ. Induction of platelet-derived growth factor A-chain and c-myc Gene Expressions by angiotensin II in culture rat vascular smooth muscle cells. *J Clin Invest* 83: 1419-1424, 1989.

29. Naftilan AJ, Pratt RE, Eldridge CS, Lin HL, Dzau VJ. Angiotensin II induces c-fos expression in smooth muscle via transcriptional control. *Hypertension* 13: 706-11, 198.

30. Nemecek G, Coughlin S, Handley D, Moskowitz M. Stimulation of aortic smooth muscle cell mitogenesis by serotonin. *Proc Natl Acad Sci USA* 83: 674-678, 1986.

31. Owen NE. Effect of prostaglandin E1 on DNA synthesis in vascular smooth muscle cells. *Am J Physiol* 250: C584-588.

32. Owen N, Villereal M. Lys-bradykinin stimulates Na$^+$ influx and DNA synthesis in cultured human fibroblasts. *Cell* 32: 979-985, 1983.

33. Owens G. Control of hypertrophic versus hyperplastic growth of vascular smooth muscle cells. *Am J Physiol* 26:H1755-1765, 1989.

34. Owens G. Differential effects of antihypertensive therapy on vascular smooth muscle cell hypertrophy, hyperploidy, and hyperplasia in the spontaneously hypertensive rat. *Circ Res* 56: 525-536, 1985.

35. Owens GK. Influence of blood pressure on development of aortic medial smooth muscle hypertrophy in spontaneously hypertensive rats. *Hypertension* 9: 178-87, 1987.

36. Owens GK. Growth response of aortic smooth muscle cells in hypertension. In: *Blood Vessel Changes in Hypertension: Structure and Function*, M.K.W.Lee (ed.), Boca Raton: CRC Press, pp 45-63, 1989.

37. Owens G, Loeb A, Gordon D, Thompson M. Expression of smooth muscle specific alpha-isoactin in cultured vascular smooth muscle cells: Relationship between growth and cytodifferentiation. *J Cell Biol* 102: 343-352, 1986.

38. Owens G, Rabinovitch P, Schwartz S. Smooth muscle cell hypertrophy versus hyperplasia in hypertension. *Proc Natl Acad Sci USA* 78: 7759-7763, 1981.

39. Owens G, Schwartz S. Alterations in vascular smooth muscle mass in the spontaneously hypertensive rat. Role of cellular hypertrophy, hyperploidy, and hyperplasia. *Circ Res* 51: 280-289, 1982.

40. Owens GK, Schwartz SM, McCanna M. Evaluation of medial hypertrophy in resistance vessels of spontaneously hypertensive rats. *Hypertension* 11: 198-207, 1988.

41. Rovner A, Murphy R, Owens G. Expression of smooth muscle and non-muscle myosin heavy chains in cultured vascular smooth muscle cells. *J Biol Chem* 261: 14740-14745, 1986.

42. Rozengurt E. Early signals in the mitogenic response. *Science* 234: 161-166, 1986.

43. Schwartz S, Ross R. Cellular proliferation in atherosclerosis and hypertension. *Prog Cardiovasc Dis* 26: 355-372, 1984.

44. Scott BT, Resink TJ, Baur U, Burgin M, Buhler FR. Amiloride sensitive activation of S6 kinase by angiotensin II in cultured vascular smooth muscle cells. *Biochem Biophys Res Commun* 151: 583-589, 1988.

45. Taubman MB, Berk BC, Izumo S, Tsuda T, Alexander RW, Nadal GB. Angiotensin II induces c-fos mRNA in aortic smooth muscle. Role of Ca$^{2+}$ mobilization and protein kinase C activation. *J Biol Chem* 264: 526-530, 198.

46. Trippodo N, Frohlich E. Similarities of genetic (spontaneous) hypertension - man and rat. *Circ Res* 48: 309-319, 1981.

47. Turla M, Thompson M, Corjay M, Owens G. Angiotensin II and arginine vasopressin increase smooth muscle alpha-actin and nonmuscle beta-actin mRNA as well as actin synthesis and content in cultured rat aortic smooth muscle cells. *J Cell Biol* 109: 244a, 1989.

# CALCIUM DEPENDENT REGULATION OF
# VASCULAR SMOOTH MUSCLE CONTRACTION

Robert S. Moreland, Jacqueline Cilea,
and Suzanne Moreland

Bockus Research Institute, Graduate Hospital
Philadelphia, PA 19146
and
The Bristol-Myers Squibb Pharmaceutical Research Institute
Princeton, NJ 08540

## INTRODUCTION

It is well established that an increase in cytosolic free calcium ($Ca^{2+}$) initiates contraction of smooth muscle, but precisely how this transduction occurs is a subject of considerable debate. Two significant advances in our understanding of the biochemical regulation of smooth muscle contraction, made in the mid 1970's, served as the basis for many subsequent investigations. In 1974, Bremel demonstrated that the primary $Ca^{2+}$ dependence of vertebrate smooth muscle contraction is associated with the thick filament, rather than the thin filament which is the case for striated muscle (1). Later, Sobieszek (2) and Aksoy et al. (3) independently demonstrated that the $Ca^{2+}$ sensitivity of actomyosin ATPase activity was associated with phosphorylation of the 20,000 $M_r$ myosin light chain (MLC). Phosphorylation of the MLC resulted from the activation of a $Ca^{2+}$ and calmodulin dependent enzyme, the MLC kinase (4). The demonstration of a direct correlation between MLC phosphorylation and both $Ca^{2+}$ dependent actin-activated myosin ATPase activity (2,3,5) and force development in either skinned (6,7) or intact (8,9) muscle fibers has resulted in an almost universal acceptance of this system as the primary regulator of smooth muscle contraction.

## THE LATCH STATE: TWO REGULATORY SYSTEMS IN SERIES

Although $Ca^{2+}$ dependent MLC phosphorylation is certainly a primary step in the activation of smooth muscle, we now know that the physiological control of contraction is probably more complex. A temporal component is apparently involved in the regulation of smooth muscle contraction. In 1975, Hellstrand and Johansson (10) demonstrated that the maximal velocity of shortening of rat portal vein was greater during the development of an isometric twitch than during the time of peak twitch force. Siegman et al. (11) later showed that energy utilization, in rabbit taenia coli, increases rapidly during force development, but falls to lower levels during force maintenance. In terms of MLC phosphorylation, Driska et al. (9) provided evidence that a decrease in the level of MLC phosphorylation preceded the fall in force upon relaxation of contracted medial strips of the swine carotid artery. These findings, as well as others, were explained in 1981 by Dillon *et al.* (12) in the "latch hypothesis" (Figure 1). These investigators hypothesized that, upon stimulation, the increase in activator $Ca^{2+}$ and the resultant increase in MLC phosphorylation initiates force development supported by rapidly cycling crossbridges. During periods of continued tissue stimulation, force is maintained at high levels but both isotonic shortening velocity and MLC phosphorylation levels fall to suprabasal levels. Therefore, force maintenance was suggested to be supported by slowly cycling, dephosphorylated crossbridges, termed "latchbridges".

Three basic tenets were associated with this hypothesis. These were that $Ca^{2+}$ dependent MLC phosphorylation was associated with the development of force (9,12), that upon MLC dephosphorylation the rapidly cycling crossbridges became slowly or non-cycling latchbridges and therefore the level of MLC phosphorylation regulated the rate of crossbridge cycling (13,14), and that the latchbridge was regulated by a second $Ca^{2+}$ dependent system with a higher $Ca^{2+}$ sensitivity than that for MLC phosphorylation (7,15,16). Although the suggestion that crossbridge cycling rate is regulated by the degree of MLC phosphorylation was and still is controversial (17-20), the basic concept of the latch state has received considerable acceptance. This hypothesis accounted reasonably well for the energetic (11,21), mechanical (14,15,22), and biochemical data (7,15,16,22) collected during contractions of almost all types of smooth muscle by many different stimulation protocols. However, as is the case in most scientific hypotheses, new experiments and new thoughts lead to revised versions.

## THE LATCH STATE REVISED: ONE REGULATORY SYSTEM

In 1988, Hai and Murphy (23,24) published two of a series of studies developing a minimal model for the regulation of a smooth muscle

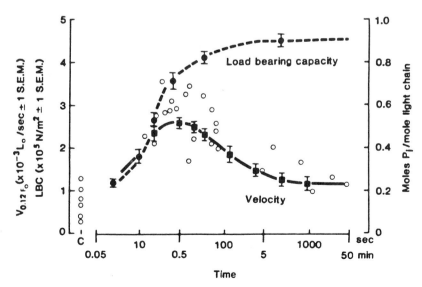

F ire 1: Log time course of a contraction of swine carotid medial fibers in re: onse to 110 mM KCl. Load-bearing capacity (●), shortening velocity at an afterload of 0.12 $F_0$ (■), and fractional MLC phosphorylation (○) are shown in this figure. Reprinted from (12) by permission (© 1981 by the AAAS).

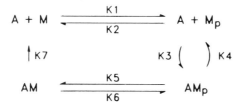

Figure 2: Kinetic model for the regulation of smooth muscle contraction. A, actin; M, detached dephosphorylated cross bridge; $M_p$, detached phosphorylated cross bridge; $AM_p$, attached phosphorylated cross bridge; and AM, attached dephosphorylated cross bridge (latch bridge). Reprinted from (23) by permission.

contraction which described the dependence of stress and crossbridge cycling rate on MLC phosphorylation. This model was composed of four states and is shown in Figure 2.

The basic assumptions for this model are: 1. A single $Ca^{2+}$ dependent regulatory system, MLC phosphorylation, is present in smooth muscle; 2. The two myosin heads are independent in terms of phosphorylation and cycling; 3. The detachment of the actin-unphosphorylated myosin complex is an irreversible process; 4. All kinetic steps can be described by first-order rate constants; 5. Phosphorylated myosin and actomyosin represent the total phosphorylated species; and 6. Phosphorylated and unphosphorylated actomyosin represent the total stress bearing species. Given these assumptions, the model shown in Figure 2 provided a remarkably close fit to the experimental data, derived primarily from diffusion-minimized or steady state experiments using the swine carotid artery (25). This approach provided a simple and testable paradigm for the $Ca^{2+}$ dependent regulation of smooth muscle. The model predicted that the relationship between stress and MLC phosphorylation was hyperbolic and that an increase in MLC phosphorylation to only 0.3 mol $P_i$/mol MLC was necessary for maximal stress development. This is an important point as it demonstrated that the model also fit data generated by other investigators who have previously shown no apparent need for large increases in MLC phosphorylation (17,26). Lastly, the model predicted a linear relationship between maximal shortening velocity and MLC phosphorylation, previously demonstrated by Murphy and his colleagues using a variety of stimulation conditions (13,27).

Although this model fit the experimental data in terms of shortening velocity, MLC phosphorylation, and stress, several important and fundamental aspects of smooth muscle physiology were not addressed by this approach. One of the most striking results predicted by this model was that 85% of the total ATP consumption by the smooth muscle cell during a contraction was due to phosphorylation by MLC kinase followed by dephosphorylation by an MLC phosphatase without crossbridge attachment (23,24), the so-called pseudo-ATPase or futile cycle. Thus, only 15% of the total energy usage was predicted to be the result of crossbridge cycling. This prediction is in contradiction to what is known concerning the energetics of smooth muscle contraction (28-30). Hellstrand and Arner (30) have examined the energetics of contraction using skinned taenia coli fibers with thiophosphorylated MLC, a condition that would significantly decrease MLC kinase and phosphatase activity and therefore the futile cycle. They found that ATP consumption during a MgATP supported contraction was depressed by approximately 15% in the absence of $Ca^{2+}$ and approximately 25% in the presence of $Ca^{2+}$; the 85% decrease in ATP consumption predicted by the Hai and Murphy model (23) was not observed. In a direct test of the model, Siegman *et al.* (31) used okadaic acid to inhibit the MLC phosphatase in the permeabilized portal vein. These investigators demonstrated that, although the $Ca^{2+}$ sensitivity of force was increased, the relationship between force and MLC phosphorylation was unaffected by okadaic acid (Figure 3). If the kinetic model proposed by Hai and Murphy (23,24) accurately reflected the steps involved in the regulation of contraction, then one would predict that inhibition of MLC phosphatase activity would, by preventing latchbridge formation upon dephosphorylation, result in a linear not a curvilinear relationship between force and MLC phosphorylation. As shown in Figure 3, this was not the case.

One of the hallmarks of smooth muscle energetics is the direct relationship between energy utilization and force. Paul and Peterson (28) and Glück *et al.* (29) clearly demonstrated that, as vascular smooth muscle is either shortened or lengthened relative to the optimal length for active force development ($L_0$), both $O_2$ consumption and force development decline (Figure 4). Unless a length dependence of MLC kinase or phosphatase activity is postulated, the model proposed by Hai and Murphy (23,24) cannot account for this linear dependence of energy utilization on force. Aksoy *et al.* (13) using intact fibers and work from our laboratory using skinned fibers of swine carotid artery (32) have shown that there are no apparent differences in either the $Ca^{2+}$ dependence (skinned), temporal relationship (intact), or magnitude (skinned and intact) of MLC phosphorylation between 0.7 and 1.4 $L_0$. It must be noted, however, that these determinations reflect net levels of MLC phosphorylation and not actual rates of MLC phosphorylation. It is therefore possible, but in our opinion not likely, that changes in length may affect MLC kinase and phosphatase activities in such a way that the final net level at any given time or $[Ca^{2+}]$ is unchanged relative to that at $L_0$.

Therefore, although the kinetic model proposed by Hai and Murphy (23,24) is exquisite in its simplicity and ability to predict several parameters of a smooth muscle contraction, it is precisely this simplicity that may

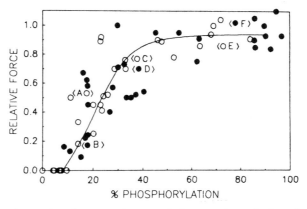

Figure 3: Relationship between force and degree of MLC phosphorylation at different $[Ca^{2+}]$ in the presence (●) and absence (○) of the phosphatase inhibitor okadaic acid (5 μM). Reprinted from (31) by permission.

account for the inability of the model to predict numerous other fundamental aspects of the regulation of a contraction. If this single $Ca^{2+}$-dependent-step model cannot account for the known experimental results concerning contractile regulation, the question remains: How does $Ca^{2+}$ contract smooth muscle?

## AN ALTERNATE HYPOTHESIS: TWO REGULATORY SYSTEMS IN PARALLEL

There is no doubt that the $Ca^{2+}$ and calmodulin dependent phosphorylation of MLC is of paramount importance in the initiation of a smooth muscle contraction. However, there is doubt as to the all inclusive role of $Ca^{2+}$ and calmodulin dependent MLC phosphorylation for the regulation of both stress and crossbridge cycling rate. Moreover, it is questionable whether MLC phosphorylation alone is necessary <u>and</u> sufficient for contraction of smooth muscle. These doubts are cast by a growing body of literature from numerous laboratories investigating the regulation of smooth muscle contraction. An increasing number of studies has presented either direct or indirect evidence for a dual role for $Ca^{2+}$ in contractile regulation: via MLC phosphorylation and via a second as yet unidentified mechanism.

An early study suggesting that calcium may directly affect crossbridge interaction was performed by Siegman *et al.* (33) who observed a calcium dependent resistance to stretch, suggestive of crossbridge attachment, in unstimulated rabbit taenia coli. Although determination of MLC phosphorylation levels was not performed for obvious reasons, under the conditions of this study an increase above basal in phosphorylation levels would not be expected. More direct approaches to the question of a second role for $Ca^{2+}$ in smooth muscle regulation were taken by Wagner and Rüegg (34) and by our laboratory (35,36). Wagner and Rüegg clearly showed that in freshly skinned chicken gizzard fibers a $Ca^{2+}$ and calmodulin dependent increase in force can be elicited without any increase in MLC phosphorylation levels. In the presence of 1.6 μM $Ca^{2+}$ and 0.05 μM

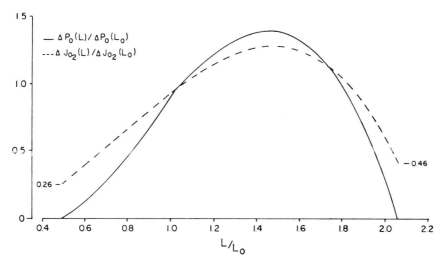

Figure 4: Dependence of active isometric force and suprabasal $O_2$ consumption rate on relative muscle length. Reprinted from (28) by permission.

calmodulin, the freshly skinned fibers developed ≈ 86% of the maximal force with no change in MLC phosphorylation (0.087 mol $P_i$/mol MLC basal; 0.087 mol $P_i$/mol MLC activated). If calmodulin were increased to 0.5 or 5.0 μM, force increased to 100% of maximal and MLC phosphorylation levels increased to 0.30 and 0.48 mol $P_i$/mol MLC, respectively. Only $Ca^{2+}$, calmodulin, and MLC phosphorylation dependent contractions could be elicited in skinned fibers stored in glycerol for longer than one week. This report suggests the possibility that the direct relationship between force and MLC phosphorylation noted by most investigators using skinned smooth muscle fibers may result from the loss of an "integral" factor during long term storage.

We have addressed the possibility of a second role for $Ca^{2+}$ using two experimental protocols, each of which was designed to ask a specific question. The first protocol was to determine if, as was shown in chicken gizzard by Wagner and Rüegg (34), detergent skinned swine carotid medial fibers can contract in the absence of MLC phosphorylation. We stimulated with either 20 mM $Mg^{2+}$ or 5 μM $Ca^{2+}$ in the presence of the ATP analog, CTP, which is a substrate for the myosin ATPase but not for the MLC kinase. As can be seen in Figure 5, both $Mg^{2+}$ and $Ca^{2+}$ produced significant levels of stress in the complete absence of any increase in the level of MLC phosphorylation (36). We have previously demonstrated, by two-dimensional electrophoresis and subsequent autoradiography of $^{32}P$ labelled tissues, that the low basal level of MLC phosphorylation shown in this figure is an artifact of the electrophoretic method and not a true measure of MLC phosphorylation (37). The second protocol was designed to determine if the $Ca^{2+}$ dependence of this apparently MLC phosphorylation independent contraction is mediated by calmodulin. For these experiments, the skinned fibers were contracted with either 20 mM $Mg^{2+}$ or 5 μM $Ca^{2+}$ in the presence of 5 mM MgCTP, allowed to reach a steady state level of stress, then exposed to increasing concentrations of the calmodulin antagonists, trifluoperazine (TFP) or W-7. As demonstrated by the results shown in Figure 6, both the $Mg^{2+}$ and $Ca^{2+}$ dependent, MgCTP supported contractions were relaxed in a concentration dependent fashion by the calmodulin antagonists. This is not a non-specific effect on the crossbridge

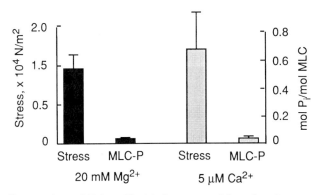

Figure 5: Contraction of Triton X-100 detergent skinned swine carotid medial fibers supported by 5 mM MgCTP in response to either $Mg^{2+}$ or $Ca^{2+}$. Reprinted from (36) by permission.

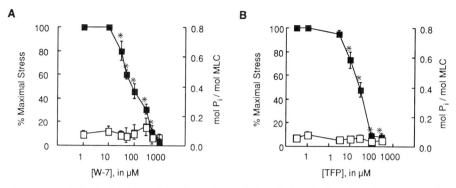

Figure 6: Effect of W-7 and TFP on stress (■) and MLC phosphorylation (□) of Triton X-100 detergent skinned swine carotid medial fibers in response to A. 5 μM $Ca^{2+}$ and 5 mM MgCTP or B. 20 mM $Mg^{2+}$ and 5 mM MgATP. *Significantly different from stress in the absence of calmodulin antagonist. Reprinted from (36) by permission.

interaction as neither antagonist affects contractions of thiophosphorylated skinned fibers. Moreover, normal $Ca^{2+}$ and MgATP dependent contractions can be elicited following washout of the calmodulin antagonists, suggesting that these agents do not inhibit contraction irreversibly. These data indicate that the divalent cation dependent, MLC phosphorylation independent contractions of smooth muscle require a calmodulin-like $Ca^{2+}$ binding protein. The results of these two experiments agree closely with those of Wagner and Rüegg (34) which also demonstrate $Ca^{2+}$ and calmodulin(-like) dependent but MLC phosphorylation independent contractions and suggest that a second $Ca^{2+}$ regulatory (or at least modulatory) system is active in smooth muscle. Moreover, this system is capable, under certain conditions, of initiating contractions of smooth muscle without prior MLC phosphorylation.

The preceding discussion has provided evidence that contraction of smooth muscle can be elicited in the absence of MLC phosphorylation. As stated in the beginning of this section however, in most circumstances, MLC phosphorylation is likely to be the predominant mechanism for the initiation of a contractile event. This secondary system is probably more important in maintaining force. Importantly, these results demonstrate that the step in the scheme proposed by Hai and Murphy (23,24) listed as K7 (Figure 2) is, in fact, reversible.

One of the most controversial roles assigned to MLC phosphorylation is that of regulating crossbridge cycling rate as determined by estimates of shortening velocity. This assignment was based on numerous studies demonstrating a linear correlation between shortening velocity and level of MLC phosphorylation (13,22,27). However, a large number of studies have provided clear evidence that such a correlation does not exist (17-20,38-41). Siegman *et al.* (17) showed that increasing extracellular calcium during electrical stimulation of rabbit taenia coli increased shortening velocity in the absence of a concomitant increase in MLC phosphorylation. Using the same tissue, Butler *et al.* (38) demonstrated, in an elegantly simple experiment, that restimulating the tissue after 20 sec of relaxation from a prior stimulation caused maximal stress redevelopment with significantly attenuated levels of both MLC phosphorylation and shortening velocity. If the muscle was restimulated after 60 sec of relaxation, force and shortening velocity increased maximally although MLC phosphorylation levels remained attenuated. One of the strongest lines of evidence suggesting that MLC phosphorylation may not be the sole regulator of shortening velocity comes from work by Gerthoffer (39,40) and Merkel *et al.* (41). Gerthoffer demonstrated that reducing extracellular calcium during carbachol stimulation of canine trachealis resulted in a decrease in both stress and shortening velocity without any decrease in the level of MLC phosphorylation (Figure 7, 39,40). A more straight forward approach was recently taken by Merkel *et al.* (41) who showed in the same tissue that stimulation with methacholine monotonically increased stress and MLC phosphorylation to sustained levels, whereas shortening velocity increased only transiently. These investigators also measured phosphorylase a activity as an indicator of cellular $[Ca^{2+}]$ and found that shortening velocity, but not MLC phosphorylation, correlated well with this index. We have performed studies, using intact swine carotid media, that also highly suggest that shortening velocity correlates with cellular calcium and not with the level of MLC phosphorylation (19). We have shown that inhibition of calcium influx through dihydropyridine-sensitive calcium channels inhibits stress maintenance and decreases shortening velocity without affecting MLC phosphorylation levels. Conversely, activation of these calcium channels with Bay k 8644 results in the slow development of stress with low (but significantly elevated) shortening velocities and no increase in MLC phosphorylation (19). Taken together, these experimental results suggest that $Ca^{2+}$ regulates, or at least modulates, crossbridge cycling rate directly and not indirectly through the level of MLC phosphorylation.

If an increase in cellular $Ca^{2+}$ can directly initiate smooth muscle contraction and can directly regulate crossbridge cycling rate independent of the MLC phosphorylation system, how can this be accomplished? Several potential $Ca^{2+}$ dependent regulatory proteins have been proposed over the past few years to account for stress maintenance or development in the face of disproportionately low levels MLC phosphorylation. These include caldesmon and the putative caldesmon kinase (42-44), calponin (45), a calmodulin-like protein (35,36), and protein kinase C (46-50). Although the evidence suggesting a role for caldesmon and calponin in smooth muscle are most intriguing, they both appear to act through a dis-inhibitory or modulatory mechanism rather than through a true activation mechanism. Protein kinase C on the other hand has been suggested to be involved in the direct activation of smooth muscle.

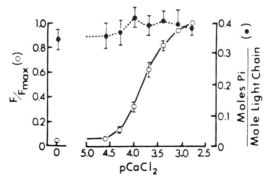

Figure 7: Calcium dependence of force and myosin phosphorylation during stimulation of tracheal muscle with carbachol. Reprinted from (39) by permission.

Chatterjee and Tajeda (48) presented evidence that phorbol esters, activators of protein kinase C, induced near maximal contraction of chemically skinned swine carotid fibers with an increase in MLC phosphorylation of only 0.08 mol $P_i$/mol MLC. Singer and Baker (48) have shown that phorbol esters induce contraction in Ca-depleted intact rabbit thoracic aortae in Ca-free solution, conditions that inhibit the MLC kinase. This finding is similar to that shown by Jiang and Morgan (49) who demonstrated that although phorbol esters contract ferret aortic strips, neither MLC phosphorylation levels nor intracellular [$Ca^{2+}$], as measured by aequorin, increase above basal. We have recently investigated the possible role protein kinase C may play in the regulation of smooth muscle contraction. Rather than activate protein kinase C with phorbol esters, we chose to inhibit protein kinase C activity with staurosporine (51). The effects of staurosporine on mechanical and biochemical parameters of a contraction were examined in intact swine carotid medial fibers stimulated with endothelin. Tissues were exposed to 80 nM staurosporine for 20 min before addition of 0.3 µM endothelin-1. Both stress and active dynamic stiffness, an estimate of the number of attached crossbridges, were determined during the endothelin induced contraction. As shown in Figure 8, endothelin produced a large well maintained increase in stress and a concomitant increase in stiffness (Figure 8A). In the presence of the protein kinase C inhibitor, stress developed to similar levels, but stiffness was significantly reduced by approximately 50% (Figure 8B). Because stiffness, indicative of the number of attached crossbridges, is decreased whereas stress is unchanged, the force per crossbridge must have increased significantly during inhibition of protein kinase C activity. This suggests protein kinase C may somehow activate a distinct population of crossbridges. These crossbridges most likely exhibit "weakly bound" characteristics because their inhibition has little effect on the magnitude of developed stress (52). The concentration of staurosporine used in this study (80 nM) was shown to inhibit a maximal phorbol ester induced contraction (protein kinase C dependent) of this tissue by > 80% with little or no effect on a contraction in response to 110 mM KCl (protein kinase C independent). Therefore, we believe these effects of staurosporine were selective for protein kinase C. This is further substantiated by the finding that there were no differences in the level or time course of MLC phosphorylation

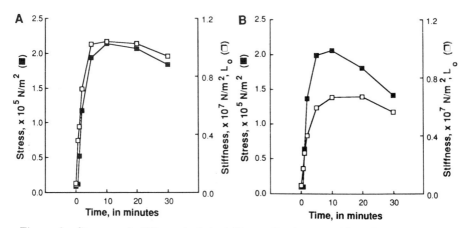

Figure 8: Stress and stiffness in intact fibers of swine carotid media stimulated with 0.3 µM endothelin in the absence (Panel A) and presence (Panel B) of 80 nM staurosporine.

when comparing the endothelin contractions in the presence and absence of staurosporine. Therefore, the significant change in force per crossbridge following inhibition of protein kinase C activity is not the result of changes in MLC phosphorylation levels. Moreover, the change in stiffness is not due to a change in series elasticity as the relationship between a decrease in tissue length and the corresponding fractional decrease in force is unaffected by staurosporine. Although a role for protein kinase C was not invoked, the concept of an agonist dependent change in the force per crossbridge has previously been suggested by Brozovich and Morgan (53).

## SUMMARY

The experimental results discussed from our laboratory as well as from numerous other laboratories investigating the regulation of smooth muscle contraction have, in our opinion, clearly demonstrated that a simple $Ca^{2+}$ dependent switch (MLC phosphorylation) cannot completely explain all of the mechanical and energetic findings. We and others have demonstrated that stress can be developed in the complete absence of increases in MLC phosphorylation (19,34,36,48,49), that crossbridge cycling rate can be regulated independent of changes in MLC phosphorylation (17-20,36-41), that $Ca^{2+}$ can directly influence both stress and crossbridge cycling rate (17,19,39-41), and that protein kinase C can, apparently, directly initiate the development of stress supported by a specific population of crossbridges characterized by unphosphorylated MLC, low cycling rates, and weak binding characteristics. This information combined with the wealth of material demonstrating the important function played by the $Ca^{2+}$ and calmodulin dependent MLC kinase is consistent with the hypothesis that there are two $Ca^{2+}$ dependent regulatory systems acting in parallel in smooth muscle (Figure 9). One of these is the $Ca^{2+}$ dependent MLC phosphorylation-dephosphorylation system responsible for the rapid development of stress and the second is a hypothesized $Ca^{2+}$ dependent system responsible for the slow development of stress as well as the maintenance of previously developed stress. This second system has a

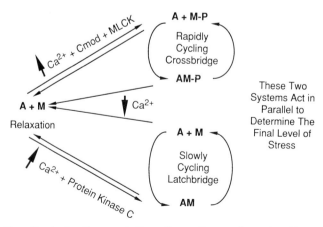

Figure 9: Hypothesis for the regulation of smooth muscle contraction. Two $Ca^{2+}$ dependent regulatory systems, MLC phosphorylation and protein kinase C, acting in parallel to determine the final stress developed and maintained.

higher $Ca^{2+}$ sensitivity than that for MLC phosphorylation and may be activated by protein kinase C. The total stress attained by smooth muscle is the result of these two regulatory systems acting in concert.

Although we believe the available information is consistent with this hypothesis of two regulatory systems functioning in parallel, it is by no means the only possibility. Early work from our laboratory (54) and the recent work by the Somlyos and their colleagues (55,56) and Kubota *et al.* (57) suggest the possibility of a regulated MLC phosphatase which might functionally alter the $Ca^{2+}$ sensitivity of the contractile filaments. Kerrick and Hoar (58) and Nishimura and van Breemen (59) have published data which imply a role for MgADP in latchbridge kinetics. These findings, as well as the discovery of several thin filament protein components (42-45) which have been proposed as regulatory units, must all be taken into account in the final answer to the question: How does $Ca^{2+}$ contract smooth muscle?

## ACKNOWLEDGMENTS

This work was supported, in part, by funds from NIH HL 37956 and from the Southeastern Pennsylvania Affiliate of the American Heart Association.

## REFERENCES

1. Bremel RD. Myosin linked calcium regulation in vertebrate smooth muscle. *Nature* 252: 405, 1974.
2. Sobieszek A. $Ca^{2+}$-linked phosphorylation of a light chain in vertebrate smooth muscle myosin. *Eur J Biochem* 73: 477, 1977.
3. Aksoy MO, Williams D, Sharkey EM, Hartshorne DJ. A relationship between $Ca^{2+}$ sensitivity and phosphorylation of gizzard actomyosin. *Biochem Biophys Res Commun* 69: 35, 1976.
4. Dabrowska R, Sherry JMF, Aromatorio DK, Hartshorne DJ. Modulator protein as a component of the myosin light chain kinase from chicken gizzard. *Biochemistry* 17: 253, 1978.
5. DiSalvo J, Gruenstein E, Silver P. $Ca^{2+}$-dependent phosphorylation of bovine aortic actomyosin. *Proc Soc Exp Med Biol* 158: 410, 1978.
6. Kerrick WGL, Hoar PE, Cassidy PS. Calcium activated tension: The role of myosin light chain phosphorylation. *Fed Proc* 39: 1558, 1980.
7. Chatterjee M, Murphy RA. Calcium dependent stress maintenance without myosin phosphorylation in skinned smooth muscle. *Science* 211: 495, 1981.
8. Barron JT, Bárány M, Bárány K, Storti RV. Reversible phosphorylation and dephosphorylation of the 20000 dalton light chain of myosin during the contraction-relaxation-contraction cycle of arterial smooth muscle. *J Biol Chem* 255: 6238, 1980.
9. Driska SP, Aksoy MO, Murphy RA. Myosin light chain phosphorylation associated with contraction in arterial smooth muscle. *Am J Physiol* 240: C222, 1981.
10. Hellstrand P, Johansson B. The force velocity relation in phasic contraction of venous smooth muscle. *Acta Physiol Scand* 93: 157, 1975.

11. Siegman MJ, Butler TM, Mooers SU, Davies RE. Chemical energetics of force development, force maintenance, and relaxation in mammalian smooth muscle. *J Gen Physiol* 76: 609, 1980.

12. Dillon PF, Aksoy MO, Driska SP, Murphy RA. Myosin phosphorylation and the cross-bridge cycle in arterial smooth muscle. *Science* 211: 495, 1981.

13. Aksoy MO, Murphy RA, Kamm KE. Role of $Ca^{2+}$ and myosin light chain phosphorylation in regulation of smooth muscle. *Am J Physiol* 242: C109, 1982.

14. Dillon PF Murphy RA. Tonic force maintenance with reduced shortening velocity in arterial smooth muscle. *Am J Physiol* 242: C102, 1982.

15. Aksoy MO, Mras S, Kamm KE, Murphy RA. $Ca^{2+}$, cAMP, and changes in myosin phosphorylation during contraction of smooth muscle. *Am J Physiol* 245: C255, 1983.

16. Moreland RS, Murphy RA. Determinants of $Ca^{2+}$-dependent stress maintenance in skinned swine carotid media. *Am J Physiol* 251: C892, 1986.

17. Siegman MJ, Butler TM, Mooers SU, Michalek A. $Ca^{2+}$ can affect $V_{max}$ without changes in myosin light chain phosphorylation in smooth muscle. *Pflügers Arch* 401: 385, 1984.

18. Haeberle JR, Hott JW, Hathaway DR. Regulation of isometric force and isotonic shortening velocity by phosphorylation of the 20,000 dalton myosin light chain of rat uterine smooth muscle. *Pflügers Arch* 403: 215, 1985.

19. Moreland S, Moreland RS. Effects of dihydropyridines on stress, myosin phosphorylation, and $V_o$ in smooth muscle. *Am J Physiol* 252: H1049, 1987.

20. Moreland S, Moreland RS, Singer HS. Apparent dissociation of myosin light chain phosphorylation and maximal velocity of shortening in KCl depolarized swine carotid artery: Effects of temperature and [KCl]. *Pflügers Arch* 408: 139, 1987.

21. Paul RJ. Coordination of metabolism and contractility in vascular smooth muscle. *Federation Proc* 42: 62, 1983.

22. Kamm KE, Stull JT. Myosin phosphorylation, force, and maximal shortening velocity in neurally stimulated tracheal smooth muscle. *Am J Physiol* 249: C238, 1985.

23. Hai CH, Murphy RA. Cross-bridge phosphorylation and regulation of latch state in smooth muscle. *Am J Physiol* 254: C99, 1988.

24. Hai CH, Murphy RA. Regulation of shortening velocity by cross-bridge phosphorylation in smooth muscle. *Am J Physiol* 255: C86, 1988.

25. Singer HS, Murphy RA. Maximal rates of activation in electrically stimulated swine carotid media. *Circ Res* 60: 438, 1987.

26. Hoar PE, Pato MD, Kerrick WGL. Myosin light chain phosphatase: Effect on the activation and relaxation of gizzard smooth muscle skinned fibers. *J Biol Chem* 260: 8760, 1985.

27. Rembold CM, Murphy RA. Myoplasmic [$Ca^{2+}$] determines myosin phosphorylation and isometric force in agonist-stimulated swine arterial smooth muscle. *J Cardiovasc Pharmacol* 12(Suppl 5): S38, 1988.

28. Paul RJ, Peterson JW. Relation between length, isometric force and oxygen consumption rate in vascular smooth muscle. *Am J Physiol* 228: 915, 1975.

29.  Glück E, Paul RJ. The aerobic metabolism of porcine carotid artery and its relation to force. Energy cost of isometric contraction. *Pflügers Arch* 370: 9, 1977.
30.  Hellstrand P, Anders A. Myosin light chain phosphorylation and the cross-bridge cycle at low substrate concentration in chemically skinned guinea pig Taenia coli. *Pflügers Arch* 405: 323, 1985.
31.  Siegman MJ, Butler TM, Mooers SU. Phosphatase inhibition with okadaic acid does not alter the relationship between force and myosin light chain phosphorylation in permeabilized smooth muscle. *Biochem Biophys Res Commun* 161: 838, 1989.
32.  Moreland RS, Moreland S, Murphy RA. Effects of length, $Ca^{2+}$, and myosin phosphorylation on stress generation in smooth muscle. *Am J Physiol* 255: C473, 1988
33.  Siegman MJ, Butler TM, Mooers SU, Davies RE. Calcium-dependent resistance to stretch and stress relaxation in resting smooth muscle. *Am J Physiol* 231: 1501, 1976.
34.  Wagner J, Rüegg JC. Skinned smooth muscle: calcium-calmodulin activation independent of myosin phosphorylation. *Pflügers Arch* 407: 569, 1986.
35.  Moreland S, Little DK, Moreland RS. Calmodulin antagonists inhibit latchbridges in detergent skinned arterial fibers. *Am J Physiol* 252: C523, 1987.
36.  Moreland RS, Moreland S. Regulation by calcium and myosin phosphorylation of the development and maintenance of stress in vascular smooth muscle. In: *Progress in Clinical and Biological Research Volume 315: Muscle Energetics*, edited by Paul RJ, Elzinga G, Yamada K. New York: AR Liss, Inc., p. 317, 1989.
37.  Moreland RS, Moreland S. Characterization of $Mg^{2+}$ induced contractions in detergent skinned swine carotid media. *Am J Physiol*, In Press.
38.  Butler TJ, Siegman MJ, Mooers SU. Slowing of cross-bridge cycling in smooth muscle without evidence of an internal load. *Am J Physiol* 251: C945, 1986.
39.  Gerthoffer WT. Calcium dependence of myosin phosphorylation and airway smooth muscle contraction and relaxation. *Am J Physiol* 250: C597, 1986.
40.  Gerthoffer WT. Dissociation of myosin phosphorylation and active tension during muscarinic stimulation of tracheal smooth muscle. *J Pharmacol Exp Ther* 2450: 8, 1987.
41.  Merkel L, Gerthoffer WT, Torphy TJ. Dissociation between myosin phosphorylation and shortening velocity in canine trachea. *Am J Physiol* 258: C524, 1990.
42.  Ngai PK, Walsh MP. Inhibition of smooth muscle actin-activated myosin $Mg^{2+}$-ATPase activity by caldesmon. *J Biol Chem* 259: 13656, 1984.
43.  Ngai PK, Walsh MP. The effects of phosphorylation of smooth-muscle caldesmon. *Biochem J* 244: 417, 1987.
44.  Adam LP, Haeberle JR, Hathaway DR. Phosphorylation of caldesmon in arterial smooth muscle. *J Biol Chem* 264: 7698, 1989.
45.  Takahashi K, Hiwada K, Kokubu T. Vascular smooth muscle calponin. A novel troponin T-like protein. *Hypertension* 11: 620, 1988.
46.  Park S, Rasmussen H. Activation of tracheal smooth muscle contraction: Synergism between $Ca^{2+}$ and activators of protein kinase C. *Proc Nat'l Acad Sci USA* 82: 8835, 1985.

47. Nishimura J, van Breemen C. Direct regulation of smooth muscle contractile elements by second messengers. *Biochem Biophys Res Commun* 163: 929, 1989.

48. Chatterjee M, Tajeda M. Phorbol ester-induced contraction in chemically skinned vascular smooth muscle. *Am J Physiol* 251: C356, 1986.

49. Singer HA, Baker KM. Calcium dependence of phorbol 12,13-dibutyrate-induced force and myosin light chain phosphorylation in arterial smooth muscle. *J Pharmacol Exp Ther* 243: 814, 1987.

50. Jiang MJ, Morgan KG. Intracellular calcium levels in phorbol ester-induced contraction of vascular muscle. *Am J Physiol* 253: H1365, 1987.

51. Nakadate T, Jeng AY, Blumberg PM. Comparison of protein kinase C functional assays to clarify mechanisms of inhibitor action. *Biochem Pharmacol* 37: 1541, 1988.

52. Warshaw DM, Desrosiers JM, Work SS, Trybus KM. Mechanical interaction of smooth muscle crossbridges modulates actin filament velocity in vitro. In: *Progress in Clinical and Biological Research Volume 327: Frontiers in Smooth Muscle Research*, edited by Sperelakis N, Wood JD. New York: Wiley-Liss, Inc., p. 815, 1990.

53. Brozovich FV, Morgan KG. Stimulus-specific changes in mechanical properties of vascular smooth muscle. *Am J Physiol* 257: H1573, 1989.

54. Moreland RS, Ford GD. The influence of $Mg^{2+}$ on the phosphorylation and dephosphorylation of myosin by an actomyosin preparation from vascular smooth muscle. *Biochem Biophys Res Commun* 106: 652, 1982.

55. Kitazawa T, Kobayashi S, Horiuti K, Somlyo AV, Somlyo AP. Receptor-coupled, permeabilized smooth muscle. Role of the phosphatidylinositol cascade, G-proteins, and modulation of the contractile response to $Ca^{2+}$. *J Biol Chem* 264: 5339, 1989.

56. Somlyo AP, Kitazawa T, Himpens B, Matthijs G, Horiuti K, Kobayashi S, Goldman YE, Somlyo AV. Modulation of $Ca^{2+}$-sensitivity and of time course of contraction in smooth muscle: A major role of protein phosphatases? In: *Advances in Protein Phosphatases, Vol. 5*, edited by: W. Merievede and J. DiSalvo, Leuven: Leuven University Press, p. 181, 1989.

57. Kubota Y, Kamm KE, Stull JT. Mechanism of GTPγS-dependent regulation of smooth muscle contraction. *Biophys J* 57: 163a, 1990.

58. Kerrick WGL, Hoar PE. Non-$Ca^{2+}$-activated contraction in smooth muscle. In: *Progress in Clinical and Biological Research Volume 245: Regulation and Contraction of Smooth Muscle*, edited by Siegman MJ, Somlyo AP, Stephens NL, New York: A.R. Liss, Inc., p. 437, 1987.

59. Nishimura J, van Breemen C. Possible involvement of actomyosin ADP complex in regulation of $Ca^{2+}$ sensitivity in α-toxin permeabilized smooth muscle. *Biochem Biophys Res Commun* 165: 408, 1989.

# CALCIUM-REGULATED PROTEIN KINASES AND LOW $K_m$ cGMP PHOSPHODIESTERASES: TARGETS FOR NOVEL ANTIHYPERTENSIVE THERAPY

Paul J. Silver, Edward D. Pagani, Wayne R. Cumiskey, Ronald L. Dundore, Alex L. Harris, King C. Lee, Alan M. Ezrin and R. Allan Buchholz

Department of Cardiovascular Pharmacology
Sterling Research Group
Rensselaer, NY 12144

## INTRODUCTION

The recent gain in knowledge over the last ten years on the intracellular mechanisms which regulate vascular smooth muscle tone has expanded opportunities for the potential discovery of novel vasodilator/antihypertensive agents. This review focuses on three intracellular enzyme systems: myosin light chain kinase (MLCK) and protein kinase C (PKC), which are $Ca^{2+}$-regulated protein kinases implicated in the control of smooth muscle tone, and the cGMP phosphodiesterases (PDEs), which regulate the levels of cGMP in smooth muscle (Fig. 1).

### Involvement of MLCK and PKC in the Regulation of Vascular Tone

Contraction of vascular and other smooth muscles is regulated by the concentration of $Ca^{2+}$ in the vicinity of the contractile proteins. During excitation, intracellular levels of $Ca^{2+}$ increase and $Ca^{2+}$ binds to calmodulin. $Ca^{2+}$ binding to calmodulin leads to $Ca^{2+}$-calmodulin mediated activation of a specific protein kinase, MLCK, which catalyzes the phosphorylation of serine-19 on the 20,000 dalton MLC. When this MLC is phosphorylated, an increase in the rate of cross bridge cycling and myosin ATPase activity occurs. Numerous MLC phosphatases have also been identified which dephosphorylate the MLC and thus reverse the effects of MLCK.

Substantial biochemical and physiological evidence supports the role of MLC phosphorylation in initiating vascular smooth muscle contraction (See refs. 1-3 for some reviews). However, it is also evident that MLC phosphorylation is not necessary for maintaining isometric force in intact vascular smooth muscle. Several additional mechanisms, including the "latch bridge" state (4), and regulation through leiotonin (5), or caldesmon (6,7), may also be involved in maintenance of smooth muscle tone. Since

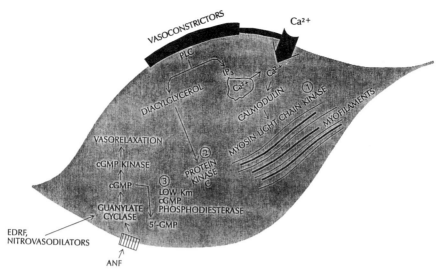

Figure 1: Sites for intracellular pharmacological modulation in vascular smooth muscle. Inhibitors of myosin light chain kinase would prevent activation of the myofilaments by $Ca^{2+}$. A second site for intervention would be inhibition of protein kinase C, which plays an important role in force maintenance. A third site would be inhibition of low $K_m$ cGMP phospho-diesterases. These inhibitors will potentiate the activity of endothelial derived relaxant factor (EDRF), nitrovasodilators, and atrial natriuretic factor (ANF).

intracellular levels of free $Ca^{2+}$ decline during the maintenance of tone (8), additional regulatory mechanisms must be sensitive to lower concentrations of free $Ca^{2+}$.

One proposed mechanism of smooth muscle tone maintenance that meets this requirement is PKC activation. Evidence supporting this hypothesis comes from studies which demonstrate sustained diacylglycerol formation by contractile agonists in intact smooth muscle cells or tissue, the ability of phorbol esters which activate PKC to produce slow, sustained contractions in various vascular smooth muscle preparations, and from experiments that have identified putative unique phosphoprotein substrates for PKC in vascular or tracheal smooth muscle (9-12). Protein kinase C activity has been found in particulate and soluble fractions of various vascular beds, with 4-6x greater activity present in particulate fractions (13). There is also a 2x greater increase in vascular reactivity to phorbol esters in vascular smooth muscle isolated from SHR relative to WKY rats, although a direct link to altered PKC activity has yet to be established.

### Pharmacological Modulation of MLCK and PKC

There are no current anti-hypertensive agents in clinical use that function solely via inhibition of vascular contractile protein interactions. Thus, these targets remain as potential sites of action for the discovery and development of novel anti-hypertensive agents.

Direct pharmacological regulation of vascular contractile protein interactions can theoretically occur at several sites. Among these are the aforementioned sites of regulation on the thin filament (caldesmon) or on the thick filament (phosphorylation of MLC). Most reported efforts to date have focused on developing modulators of MLC phosphorylation as an approach to pharmacologically regulate vascular contractile protein interactions. Direct alteration of MLC phosphorylation can conceivably occur by four distinct modes including inhibition of $Ca^{2+}$ binding to calmodulin, inhibition of $Ca^{2+}$-calmodulin activation of MLCK, direct inhibition of MLCK catalytic activity, or direct stimulation of MLC phosphatase activity. Organic agents which relax vascular smooth muscle and function exclusively by stimulating phosphatase activity have not been identified. The knowledge that the mechanism of regulation of MLCK, like other protein kinases, involves pseudosubstrate inhibition (14) may aid in the design of specific MLCK inhibitors. Efforts of two pharmaceutical groups over the past few years have resulted in some small peptide substrate inhibitors of MLCK, with $K_i$s mostly in the low micromolar or nanomolar range (15,16). However, because of the peptide nature of these agents and the need for access to intracellular contractile proteins, functional inhibition of vascular force development is not obtained.

Calmodulin antagonism is currently the most popular method for inhibiting smooth muscle MLC phosphorylation. By far, the largest group of calmodulin antagonists are those agents which compete with the regulated enzyme for the $Ca^{2+}$-calmodulin complex. Agents which directly modify $Ca^{2+}$ binding to calmodulin, such as HT-74, are few (17). Many classes of drugs have been shown to bind to calmodulin in a $Ca^{2+}$-dependent manner, and also to inhibit various calmodulin-regulated systems. Among these are phenothiazine, diphenylbutylpiperidine and butyrophenone antipsychotics, tricyclic and nontricylic antidepressants, insect venoms (such as mastoparan and melittin), benzodiazepine antianxiety agents, naphthalene sulfonamide smooth muscle relaxants, the antifungal miconazole analog calmidazolium, class III antiarrhythmic agents such as amiodarone, and certain lipophilic $Ca^{2+}$ entry blockers (18-23).

The potency of calmodulin inhibitors in most biochemical and skinned muscle preparations generally ranges from 1 to 30 µM. Most vasodilators (such as nitroprusside, hydralazine and diazoxide), and certain $Ca^{2+}$ entry blockers (nifedipine and verapamil) do not affect contractile proteins at concentrations as high as 100 µM (24). Some dihydropyridine $Ca^{2+}$ blockers (felodipine) as well as weakly basic lipophilic $Ca^{2+}$ antagonists (such as cinnarizine, flunarizine, bepridil, and perhexiline) also bind to calmodulin and inhibit calmodulin-regulated MLC phosphorylation and phosphodiesterase activity (22,25). It is doubtful that this activity contributes to their pharmacological activity, since potency for $Ca^{2+}$ entry blockade is generally several orders of magnitude higher for most of these agents.

The pharmacological effects of two novel series of agents which structurally resemble the dihydropyridines or verapamil/W-7 hybrids have been described. Both series demonstrate $Ca^{2+}$ entry blockade and direct inhibition of aortic actin-myosin interactions at a similar concentration range (3-30 µM). These agents can be differentiated from standard $Ca^{2+}$ entry blockers in intact muscle or cellular preparations (26). That is, these agents are more efficacious versus most mediators (norepinephrine, angiotensin II, histamine, serotonin, leukotrienes, $PGF_{2\alpha}$ thromboxane $A_2$) of vasoconstriction in rabbit aortic or porcine coronary arterial smooth muscle. These effects are consistent with an intracellular effect at the

TABLE 1

## COMPARATIVE EFFECTS OF STAUROSPORINE AND H7 ON PKC AND MLCK ACTIVITY, AND VASORELAXATION

| | PKC Inhibition $(K_i)$[a] | MLCK Inhibition $(K_i)$[a] | Vasorelaxation $(EC_{50})$[b] | |
| --- | --- | --- | --- | --- |
| | | | Phenylephrine | Phorbol Dibutyrate |
| Staurosporine | 5 nM | 21 nM | 75 nM | 365 nM |
| H7 | 28 μM | 120 μM | 2 μM | 125 μM |

a   $K_i$ values relative to [ATP] were determined for purified PKC (rat brain) and MLCK (bovine tracheal smooth muscle). Both staurosporine and H7 are apparent competitive inhibitions of both kinases.

b   $EC_{50}$ values are the estimated concentration of staurosporine or H7 which produced 50% relaxation of phenylephrine- or phorbol dibutyrate-mediated contractions of intact SHR aortic smooth muscle strips. Developed force prior to addition of either inhibitor was 1.7 ± 0.1 g (phenylephrine) and 2.2 ± 0.1 g (phorbol dibutyrate).

contractile proteins, since vasoconstrictor agents may mobilize pools of $Ca^{2+}$ (intracellular and/or entry through voltage-independent $Ca^{2+}$ channels) which are not amenable to inhibition by standard $Ca^{2+}$ entry blockers. These agents are also effective at inhibiting platelet aggregation, which is also regulated by protein phosphorylation. However, while these agents lowered blood pressure in conscious SHR, it was not possible to discern an intracellular effect from that of $Ca^{2+}$ entry blockade.

To further delineate the role of MLCK and/or PKC in blood pressure regulation, we have examined the relationships between MLCK and PKC inhibition, vasorelaxation and blood pressure regulation in SHR using putative protein kinase inhibitors (28,29) of varying potency in biochemical, vascular tissue and *in vivo* models. Protein kinase inhibitors from two chemical classes, staurosporine-like (staurosporine, K252a) and isoquinoline sulfonamides (H7, HA1004) were tested for their ability to inhibit PKC and MLCK activity, produce vascular relaxation in phenylephrine-contracted aortic smooth muscle, and lower arterial pressure in conscious SHR.

All compounds inhibited both kinases, with an approximate 4x increase in potency for PKC relative to MLCK (Table 1). The potency for inhibition of PKC or MLCK from SHR homogenates by these inhibitors was similar to the potency obtained for enzymes purified from rat brain (PKC) or bovine trachealis (MLCK). Staurosporine produced concentration-dependent relaxation of phenylephrine-contracted SHR aorta, with an $EC_{50}$ value of 75 nM. The vasorelaxation curve for staurosporine fell between and overlapped the concentration-response curves for both PKC and MLCK inhibition. H7 also caused concentration-dependent relaxation of aortic smooth muscle ($EC_{50}$=2 μM), but was much less potent than staurosporine. In contrast to staurosporine, the vasorelaxation curve for H7 was clearly positioned to the left of the PKC and MLCK inhibition curves. HA1004 also exhibited a similar pattern, i.e., the curve depicting vasorelaxation was to the left of the inhibition curves for both PKC and MLCK. These data suggest that vasorelaxation by staurosporine, but not H7 or HA1004, may be related to inhibition of PKC and/or MLCK. Interestingly, staurosporine and H7 also relaxed phorbol dibutyrate-induced contractions in this model, with a 4-5 fold reduced potency for staurosporine ($EC_{50}$ = 365 nM) relative to phenylephrine, and a 60-fold reduced potency for H7 ($EC_{50}$ = 120 μM).

In conscious SHR, the calcium channel blocker nitrendipine and staurosporine produced dose-dependent reductions in MAP with comparable potency, as indicated by similar $ED_{25}$ values (0.24-0.28 mg/kg). However, staurosporine exhibited a much steeper dose-response relationship for lowering MAP than nitrendipine. HA1004 was the only other compound that produced at least a 25% reduction in MAP. K252A caused only a slight reduction in MAP at the highest dose tested. Higher doses were not tested due to the limits of solubility for K252A. H7 failed to reduce MAP at lower doses and produced seizures upon administration of 6 mg/kg, the highest dose tested.

In further studies, both staurosporine and nitrendipine lowered MAP significantly for the first hour after administration of equihypotensive doses. However, while MAP returned to control levels within 2 hours after injection of nitrendipine, MAP was significantly reduced for the entire 10 hour recording period after injection of staurosporine. Both staurosporine and nitrendipine produced tachycardia which was temporally related to decreases in MAP.

Oral administration of staurosporine (1.8 mg/kg) produced a gradual reduction in the MAP of conscious SHR. The maximum reduction in MAP occurred 6 hours after medication. A slight but nonsignificant increase in heart rate was observed in both the vehicle and staurosporine treated rats. This effect was probably due to the handling associated with oral drug administration. The absence of the persistent tachycardia that was observed after intravenous injection of staurosporine is probably related to the very gradual fall in arterial pressure noted after oral administration of this agent.

In further studies, the hemodynamic responses to intravenous staurosporine or nitrendipine were compared in conscious chronically-instrumented normotensive dogs. Both agents produced comparable dose-dependent reductions in MAP, with a maximum reduction of 24% at 100 µg/kg. The decreases in MAP were accompanied by significant reductions in systemic vascular resistance (SVR), although nitrendipine caused significantly greater reductions at 3-100 µg/kg. The reduction in MAP and SVR caused by nitrendipine were associated with dose-dependent increases in heart rate, cardiac output and left ventricular dP/dt, and no change in renal blood flow. In marked contrast, the fall in MAP caused by staurosporine produced less tachycardia, only small increases in cardiac output and no change in left ventricular dP/dt. In addition, staurosporine significantly increased renal blood flow, with a 50% increase at 100 µg/kg. These data indicate that both agents are equieffective at lowering arterial pressure but have significantly different hemodynamic profiles. The greater reduction in SVR caused by nitrendipine is compensated by the marked increases in myocardial contractility, heart rate and cardiac output that most likely are reflexive in nature. These effects were not observed with staurosporine. Moreover, staurosporine, but not nitrendipine, produced highly desirable increases in renal blood flow. The cardiovascular actions of staurosporine suggests that it modifies reflex sympathetic activity by an unknown mechanism, resulting in a hemodynamic profile that would be highly desirable in an antihypertensive agent.

*In toto*, these *in vivo* studies suggest that a protein kinase inhibitor like staurosporine is efficacious in at least two species and may offer some potential advantages over other antihypertensives such as $Ca^{2+}$ entry blockers. Interestingly, there were no apparent side-effects with staurosporine in either conscious rats or conscious dogs, which is surprising given the ubiquitous distribution of PKC. A major question which is still unresolved from these studies centers on the need for inhibition of MLCK, PKC or both kinases, to obtain full antihypertensive efficacy.

The importance of MLC phorphorylation vs. other contractile protein regulatory mechanisms such as PKC activation in the chronic regulation of blood pressure is still unknown. It is difficult to know the extent of MLC phosphorylation in arterioles that are ultimately responsible for regulation of vascular resistance. If MLC phosphorylation is transient and this transient response is dependent on the contractile agonist/vasoconstrictor (as it is in large arteries), then regulatory mechanisms that are more crucial to maintenance of tone (i.e., PKC) may be better targets for antihypertensive therapy. MLCK inhibitors may be better for smooth muscle diseases which involve phasic contractions, such as angina. The answer to this question awaits the development of more potent, selective inhibitors of MLCK and PKC.

## LOW $K_m$ cGMP PDE

Both low $K_m$ cAMP and low $K_m$ cGMP PDE activity are present in most, if not all, mammalian vascular smooth muscles (30). The cAMP selective enzyme has a low $K_m$ for cAMP (usually 0.2-0.4 µM) and activity can be modulated by cGMP, which can inhibit this PDE at submicromolar concentrations. This vascular PDE isozyme can be selectively inhibited by agents such as milrinone, piroximone, imazodan, CI-930, trequinsin and others in a variety of species, including primates (31). Inhibition of this PDE is accompanied by increases in cAMP content, activation of cAMP protein kinase, vasorelaxation and a reduction in vascular resistance in a manner analogous to what occurs with other cAMP-related vasodilators (32,33). Selective inhibitors of this PDE also inhibit the same isozyme(s) in cardiac muscle and ultimately increase the rate and force of contraction in the heart (30). Thus, the hemodynamic benefits of these agents in the acute therapy of heart failure may derive from a single mechanism, low $K_m$ cAMP PDE inhibition, which produces peripheral vasodilation (afterload reduction), coronary vasodilation, and positive inotropy. However, as the same isozyme is present in cardiac muscle, this is not a viable target in vascular smooth muscle for antihypertensive therapy, since the accompanying increases in positive inotropy via inhibition of this PDE in the heart would be contraindicated in the therapy of hypertension. This may have been a factor in the discontinuance of the development of the potent PDE inhibitor, trequinsin, for hypertension.

Another PDE isozyme which hydrolyzes cAMP and is not inhibited by cGMP has also been identified in vascular smooth muscle, as well as tracheal smooth muscle (34,35). This isozyme is sensitive to inhibition by low micromolar concentrations of rolipram and RO 20-1724. Unlike the cGMP-inhibitable low $K_m$ cAMP PDE isozyme previously described, the role of this PDE in regulating vascular relaxation has yet to be determined, although rolipram also reduces blood pressure in SHR.

Two forms of low $K_m$ cGMP PDE are present in vascular smooth muscle from a variety of species (36), including primates (Fig. 2). Both isozymes have $K_m$'s for cGMP in the 0.2-1 µM range. One isozyme, which is similar to isozymes present in other tissues (including brain and heart),

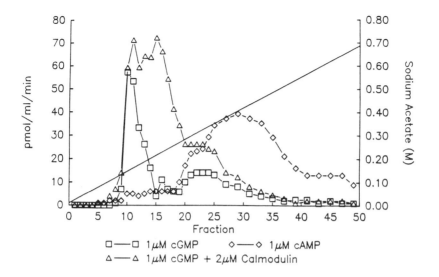

Figure 2: Phosphodiesterase (PDE) isozymes present in monkey aortic smooth muscle were separated by DEAE-Sephacel chromatography using a sodium acetate gradient as described in the text (top panel). PDE activity in each fraction was assayed in the presence of 1 µM cGMP, 1 µM cAMP or 1 µM cGMP with 2 µM calmodulin. The chemical structure of zaprinast (M & B 22948), a selective inhibitor of the first peak of low $K_m$ cGMP PDE, along with the chemical structure of cGMP, are shown. The first peak of cGMP PDE activity has a $K_m = 0.2$ µM and is inhibited by zaprinast ($IC_{50} = 0.5$ µM). The second peak of cGMP PDE activity is sensitive to activation by $Ca^{2+}$-calmodulin and has a lower affinity for cGMP ($K_m = 1$ µM). Further purification of this PDE with a calmodulin-affinity column suggests that this enzyme belongs to the family of calmodulin-regulated PDEs present in other tissues including brain and heart.

also hydrolyzes cAMP and is activated by $Ca^{2+}$-calmodulin. Calmodulin increases $V_{max}$ and does not alter the $K_m$ for cGMP. A second isozyme is also present in most, if not all, vascular smooth muscles examined. This cGMP PDE is not regulated by calmodulin but is inhibited by a selective inhibitor, zaprinast, at low micromolar concentrations ($IC_{50} = 0.5$ µM).

We have used zaprinast in biochemical, intact tissue and *in vivo* experiments to determine the role of this PDE isozyme in the regulation of vasomotor tone. In studies comparing SHR and WKY rat and guinea pig aortic strips (37), zaprinast was equally effective in relaxing phenylephrine-contracted aortae from SHR and WKY with an intact endothelium ($EC_{50} = 8$-$9$ µM) but did little in intact and denuded phenylephrine-contracted

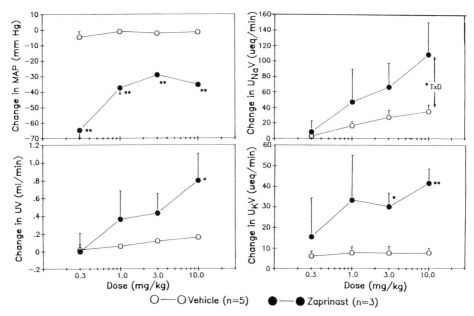

Figure 3: Changes in mean arterial pressure (MAP), urine output (UV), urinary Na excretion rate ($U_{Na}V$) and urinary K excretion rate ($U_KV$) in response to intravenous injections of zaprinast (0.3, 1.0, 3.0, 10.0 mg/kg, cumulative, 10 min between injections) or vehicle in anesthetized dogs. * = $p < 0.05$; ** $p < 0.01$ relative to vehicle-treatment when analyzed by one-way ANOVA with repeated measures. TxD indicates a significant effect due to treatment.

guinea pig aortae, or denuded SHR and WKY aortae. To test functional cGMP PDE inhibition, we assessed the ability of zaprinast to potentiate vasorelaxation by an activator of soluble (sodium nitroprusside-SNP) and particulate (atriopeptin II) guanylate cyclase. Pretreatment with zaprinast potentiated the vasorelaxant potency of SNP in both SHR and WKY aortae whereas atriopeptin II responses were potentiated only in WKY aortae. In studies with the low $K_m$ cGMP PDE isolated via DEAE column chromatography, the apparent $K_m$ for cGMP and potency of zaprinast were approximately 2 fold greater in WKY than in SHR aortic preparations. However, the $V_{max}$ for cGMP hydrolysis was greater in SHR than in WKY.

In conscious SHR (38), zaprinast (3-30 mg/kg) was injected intravenously during a steady state intravenous infusion of SNP (15 µg/kg/min). SNP significantly increased the depressor response to zaprinast. In contrast, the adenylate cyclase activator fenoldopam (20 µg/kg/min) did not affect the depressor response to zaprinast. In further studies, zaprinast (10 mg/kg) potentiated the depressor response to SNP but not atriopeptin II. While the differential effects of these two types of guanylate cyclase activators are similar to *in vitro* results with isolated aortic strips, the *in vivo* data may also reflect the different depressor mechanisms of action of SNP and atriopeptin II.

In further studies, the natriuretic and diuretic effects of zaprinast were examined in normotensive conscious rats and normotensive anesthetized dogs. In Sprague Dawley rats, zaprinast produced dose-related natriuresis from 3-30 mg/kg, p.o. No significant diuresis was noted although elevations in K+ excretion were evident over this dose range. In anesthetized dogs, zaprinast tended to increase natriuresis and kaliuresis at 3-10 mg/kg, i.v. (Fig. 3). Diuresis was also increased at 10 mg/kg. These favorable effects in renal function were evident at doses of zaprinast which produced a 30-40 mmHg decrease in MAP. While the mechanistic basis for enhanced natriuresis in these models is not known, these effects are consistent with potentiation of cGMP-mediated natriuresis in the kidney.

In summary, these data show that zaprinast potentiates the vasorelaxant and hypotensive actions of guanylate cyclase activators, including EDRF, nitrovasodilators, and ANF. In addition, zaprinast lowers blood pressure and is natriuretic in rats or dogs. Given the link between via guanylate cyclase activation and cGMP formation in the mechanisms of action of EDRF and ANF, a selective agent that prevents the degradation of cGMP in vascular and renal cells offers a new and exciting approach for the discovery of novel vasodilators that may be useful therapy for hypertension. Certainly, an agent that potentiates the activity of endogenous regulators of vascular tone and sodium balance (such as EDRF and ANF) and ultimately produces both vasodilation with accompanying natriuresis would be different from current therapeutic agents. Further developments in this hypothesis await additional information on the low $K_m$ cGMP PDE(s) in ANF-responsive renal cells, as well as the discovery of nanomolar potent/specific inhibitors of the cGMP isozyme present in vascular smooth muscle.

REFERENCES

1.  Walsh MP. Calcium regulation of smooth muscle contraction. In: *Calcium and Cell Physiology.* D. Marme (ed), Berlin: Springer-Verlag, 1985.
2.  Kamm KE, Stull JT. Regulation of smooth muscle contractile elements by second messengers. *Ann Rev Physiol* 51: 299, 1989.
3.  Moreland RS, Cilea J and Moreland S. Calcium and phosphorylation dependent regulation of vascular smooth muscle contraction. In: *Cellular and Molecular Mechanisms of Hypertension.* R.H. Cox (ed), New York: Plenum Publishers, (in press), 1990.
4.  Dillon PF, Aksoy MD, Driska SP, Murphy RA. Myosin phosphorylation and the cross-bridge cycle in arterial smooth muscle. *Science* 211: 495, 1981.
5.  Ebashi S. Regulation of contractility. In: *Muscle and Non-Muscle Motility.* A. Stracher (ed). New York: Academic Press, 1983.
6.  Ngai PK, Walsh MP. Inhibition of smooth muscle actin.activated myosin $Mg^{2+}$-ATPase by caldesmon. *J Biol Chem* 259: 13656, 1984.
7.  Smith CWS, Prichard K, Marston SB. The mechanism of $Ca^{2+}$ regulation of vascular smooth muscle thin filaments by caldesmon and calmodulin. *J Biol Chem* 262: 116, 1987.
8.  Morgan JP, Morgan KG. Stimulus-specific patterns of intracellular calcium levels in smooth muscle of ferret portal vein. *J Physiol (Lond)* 352: 155, 1984.
9.  Rasmussen H, Takuwa Y, Park S. Protein kinase C in the regulation of smooth muscle contraction. *FASEB J* 1: 177, 1987.
10. Griendling KK, Rittenhouse SE, Brock TA, et al. Sustained diacylglycerol formation from inositol phospholipids in angiotensin

II-stimulated vascular smooth muscle cells. *J Biol Chem* 261: 5901, 1986.

11. Catterjee M, Tejada M. Phorbol ester-induced contraction in chemically-skinned vascular smooth muscle. *Am J Physiol* 251: C356, 1986.

12. Parks S, Rasmussen H. Carbachol-induced protein phosphorylation changes in bovine tracheal smooth muscle. *J Biol Chem* 261: 15734, 1986.

13. Silver PJ, Lepore RE, Cumiskey WR, Kiefer D, Harris AL. Protein kinase C activity and reactivity to phorbol ester in vascular smooth muscle from spontaneously hypertensive rats (SHR) and normotensive Wistar Kyoto rats (WKY). *Biochem Biophys Res Commun* 154: 272, 1988.

14. Kemp BE, Pearson RB, Guerriero V, Bagchi IC, Means AR. The calmodulin binding domain of chicken smooth muscle myosin light chain kinase contains a pseudosubstrate sequence. *J Biol Chem* 262: 2542, 1987.

15. Moreland S, Hunt JT. Analogs of the calmodulin binding site of myosin light chain (MLC) kinase. *FASEB J* 46: 1098, 1987.

16. Foster CJ, Gaeta FCA. The calmodulin binding domain of chicken gizzard myosin light-chain kinase contains two non-overlapping active site directed inhibitory sequences. *Biophys J* 53: 182a, 1988.

17. Tanaka T, Umekuwa H, Saitoh M, *et al*. Modulation of calmodulin function and of $Ca^{2+}$-induced smooth muscle contraction by the calmodulin antagonist, HT-74. *Mol Pharmacol* 29: 264, 1986.

18. Prozialeck WC. Structure-activity relationships of calmodulin antagonists. *Ann Rep Med Chem* 18: 203, 1983.

19. Hidaka H, Tanaka T. Naphthalenesulfonamides as calmodulin antagonists. In: *Methods in Enzymology, Vol. 102, Calmodulin and Calcium-Binding Proteins*. A.R. Means and B.W. O'Malley (eds), New York, Academic Press, 1983, p 185.

20. Roufogalis BD. Calmodulin antagonism. In: *Calcium and Cell Physiology*. D. Marme (ed), Berlin: Springer-Verlag, 1985.

21. Mannold R. Calmodulin–structure, function and drug action. *Drugs of the Future* 9: 677, 1984.

22. Silver PJ, Dachiw J, Ambrose JM, Pinto PB. Effects of the calcium antagonists perhexiline and cinnarizine on vascular and cardiac contractile protein function. *J Pharmacol Exp Ther* 234: 629, 1985.

23. Silver PJ, Connell ML, Dillon KM, Cumiskey WR, Volberg WA, Ezrin AM. Inhibition of calmodulin and protein kinase C by amiodarone and other class III antiarrhythmic agents. *Cardiovasc Drugs and Therapy* 3: 657, 1989.

24. Silver PJ, Dachiw J, Ambrose JA. Effects of calcium antagonists and vasodilators on arterial myosin phosphorylation and actin-myosin interactions. *J Pharmacol Exp Ther* 230: 141, 1984.

25. Johnson JD, Fugman DA. Calcium and calmodulin antagonists binding to calmodulin and relaxation of coronary segments. *J Pharmacol Exp Ther* 226: 330, 1983.

26. Silver PJ, Sulkowski TJ, Lappe RW, Wendt RL. Wy-46-300 and Wy-46,531: Vascular smooth muscle relaxant/antihypertensive agents with combined $Ca^{2+}$ antagonist/myosin phosphorylation inhibitory mechanisms. *J Cardiovasc Pharmacol* 8: 1168, 1986.

27. Silver PJ, Fenichel R, Wendt RL. Structural variants of verapamil and W-7 with combined $Ca^{2+}$ entry blockade/myosin phosphorylation inhibitory mechanisms. *J Cardiovasc Pharmacol* 11: 299, 1988.

28. Hidaka H, Inagaki M, Kawamoto S, Sasaki Y. Isoquinoline sulfonamides, novel and potent inhibitors of cyclic nucleotide dependent protein kinase and protein kinase C. *Biochemistry* 23: 5036, 1984.

29. Tamaoki T, Nomoto H, Takahashi I, Kato Y, Morimoto M, Tomita F. Staurosporine, a potent inhibitor of phospholipid/Ca++ dependent protein kinase. *Biochem Biophys Res Commun* 135: 397, 1986.

30. Silver PJ, Pagani ED. Biochemical and pre-clinical pharmacology of selective inhibitors of cardiovascular phosphodiesterase isozymes. In: *Inotropic Drugs: Basic Research and Clinical Practice.* P Allen, J. Gwathmey, M. Briggs (eds). New York: Marcel Dekker, (in press), 1990.

31. Silver PJ, Harris AL. Phosphodiesterase isozyme inhibitors and vascular smooth muscle. In: *Proceedings of the Second International Symposium on Resistance Vessels.* W. Halpern, J. Brayden, N. McLaughlin *et al* (eds). Ithaca: Perinatology Press, 1988.

32. Kauffman RF, Scheneck KM, Utterback BG, Crowe VG, Cohen MC. *In vitro* vascular relaxation by new inotropic agents: Relationship to phosphodiesterase inhibition and cyclic nucleotides. *J Pharmacol Exp Ther* 242: 864, 1987.

33. Silver PJ, Lepore RE, O'Connor B, Lemp BM, Bentley RG, Harris AL. Inhibition of the low $K_m$ cAMP phosphodiesterase and activation of the cyclic AMP system in vascular smooth muscle by milrinone. *J Pharmacol Exp Ther* 247: 34, 1988.

34. Prigent AF, Fougier S, Nemoz G, *et al.* Comparison of cyclic nucleotide phosphodiesterase isoforms from rat heart and bovine aorta. Separation and inhibition by selective reference phosphodiesterase inhibitors. *Biochem Pharmacol* 37: 3671, 1988.

35. Silver PJ, Hamel LT, Perrone MH, Bentley RG, Bushover CR, Evans DB. Differential pharmacologic sensitivity of cyclic nucleotide phosphodiesterase isozymes isolated from cardiac muscle, arterial and airway smooth muscle. *Eur J Pharmacol* 150: 85, 1988.

36. Lugnier C, Schoeffter P, LeBec A, Strouthou E, Stoclet JC. Selective inhibition of cyclic nucleotide phosphodiesterases of human, bovine and rat aorta. *Biochem Pharmacol* 35: 1743, 1986.

37. Harris AL, Lemp BM, Bentley RG, Perrone MH, Hamel LT, Silver PJ. Phosphodiesterase isozyme inhibition and the potentiation by zaprinast of endothelium-derived relaxing factor and guanylate cyclase stimulatory agents in vascular smooth muscle. *J Pharmacol Exp Ther* 249: 394, 1989.

38. Buchholz RA, Dundore RL, Pratt PF, Hallenbeck WD, Wassey ML and Silver PJ. The selective phosphodiesterase I inhibitor zaprinast (ZAP) potentiates the hypotensive effect of sodium nitroprusside (SNP) in conscious SHR. *FASEB J* 3: A1186, 1989.

39. Lal B, Dohadwalla AN, Dadkar NK, D'sa A, de Souza NJ. Trequinsin, a potent new antihypertensive vasodilator in the series of 2-(arylimino)-3-alky1-9,10-dimethoxy-3,4,6,7-tetrahydro-2H-pyrimido [6,1-a] isoquinolin-4-ones. *J Med Chem* 27: 1470, 1984.

# MOLECULAR ASPECTS OF VOLTAGE-DEPENDENT ION CHANNELS

Robert L. Barchi

Mahoney Institute of Neurological Sciences
University of Pennsylvania School of Medicine
Philadelphia, PA 19104

## INTRODUCTION

The control of contractility in cardiac and smooth muscle ultimately depends, as it does in skeletal muscle and nerve, on electrical signals generated in the surface membrane. These signals are the result of changes in the ionic conductance of the surface membrane mediated by channel-forming transmembrane proteins. Over the past few years, concomitant with rapid developments in biochemical and molecular biological approaches to the study of these proteins, our understanding of the structure of ion channels and of the relationship between their structure and function has advanced considerably. In this chapter, current concepts of the structure of voltage-dependent ion channels will be considered with particular emphasis on the voltage-dependent sodium channel.

Voltage-dependent ion channels form a loose superfamily of proteins that share certain structural features and an appreciation of these conserved features can provide insight into the mechanisms by which these molecules carry out common actions. However, even a single type of ion channel may show minor variations in structure or function between tissues of a given species and even between regions of the same tissue (1). Knowledge of those aspects of channel structure that confer this uniqueness is also critical for an appreciation of channel function since it is often the subtle differences between related channels which can be exploited for therapeutic benefit and drug specificity. Both the common aspects and the unique features of channel structure merit attention.

## VOLTAGE-DEPENDENT SODIUM CHANNELS

The propagated action potentials characteristic of nerve and muscle are largely the result of time- and voltage-dependent changes in membrane conductance to sodium ions (2). The voltage-dependent sodium channel that controls this conductance has been the target of intensive biophysical and biochemical study since the development of the voltage clamp in the late 1940's (see ref 3 for review). The transiently activating and spontaneously inactivating sodium currents gated by this channel are remarkably similar in excitable membranes from a wide range of species throughout the phylogenetic tree, suggesting a highly refined and conserved structure for the channel protein itself.

Since the sodium channel plays such a central role in the function of nerve and of cardiac and skeletal muscle, it has become the target for a wide range of biological toxins. These toxins, in turn, have proven useful as probes of channel structure and function. At least five classes of binding sites for these toxins have been defined on pharmacological or physiological grounds (4). Of these toxins, the small polar molecules tetrodotoxin (TTX) and saxitoxin (STX) are the most familiar; these toxins block access of ions to the channel pore, bind with high affinity to the channel protein, and can be radiolabeled to high specific activity (5). Because of these properties, they have proven particularly useful in biochemical studies of the isolated channel protein as well as in separating subtypes of sodium channels on the basis of their relative binding affinities.

## COMMON ASPECTS OF SODIUM CHANNEL STRUCTURE

Using labeled TTX or STX as a marker, the sodium channel protein has been solubilized and purified from a number of excitable tissues (see ref 6 for review). In their native state, these proteins appear to be about 300,000 MW. Most of this mass is contributed by a single large subunit of 260,000 to 290,000 MW designated the alpha subunit. These alpha subunits are heavily glycosylated, containing between 26 and 30% by weight carbohydrate (7-9). In some species such as the eel, alpha is the only component present. In mammalian brain and skeletal muscle, one or more smaller beta subunits of 30-40 kDa are stoichiometrically associated with the alpha subunit (10-12). Most of these purified sodium channel proteins have been functionally reconstituted (13-15); even the purified eel channel with a single alpha subunit retains the basic selectivity and gating characteristics associated with the native channel.

cDNA encoding sodium channel alpha subunits have been cloned from eel (16), rat brain (17), rat skeletal muscle (18,19) and cardiac muscle (20), and from *Drosophila* (21,22). The mRNA transcripts encoding these alpha subunits range between 8.5 and 9.5 kb and encode proteins between 1819 and 2020 amino acids in length. Scanning these sequences immediately reveals the presence of four large internal repeat domains (16,17), each spanning about 275 amino acids. There is strong homology between species within each of these domains as well as internal homology among the domains of a single channel sequence. The level of sequence conservation implies that these internal repeat domains arose from duplication of a primordial gene early in evolution.

Hydropathicity analysis of repeat domain primary sequences indicates the presence of at least six putative transmembrane helical regions in each (17,23,24). These putative helices are the most highly conserved regions within the repeat domain. The S4 helix in each domain exhibits a repeating pattern of a positively charged residue (lysine or arginine) in every third position with nonpolar residues interspersed between (figure 1). When organized as a helix, this pattern produces a spiral band of positive charge along one face of the helix with predominantly nonpolar residues on the remaining surface. This highly charged amphipathic helix may play a primary role in voltage-dependent gating of the channel, possibly contributing to the electrophysiologically

## S - 4 HELIX
### DOMAIN IV

| | |
|---|---|
| muscle I | S P T L F R V I R L A R I G R V L R L I R G A K G I R T |
| muscle II | S P T L F R V I R L A R I G R I L R L I R G A K G I R T |
| brain I | S P T L F R V I R L A R I G R I L R L I K G A K G I R T |
| brain II | S P T L F R V I R L A R I G R I L R L I K G A K G I R T |
| brain III | S P T L F R V I R L A R I G R I L R L I K G A K G I R T |
| eel | S P T L F R V I R L A R I A R V L R L I R A A K G I R T |
| fly | S P T T L R V V R L A R I G R I L R L I K G A K G I R T |

Figure 1.

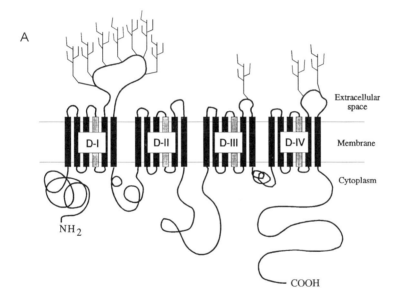

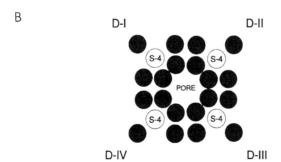

Figure 2.

observed gating charge through an axial screw-like motion normal to the plane of the membrane (25). A requirement for the S4 helix in each of the four repeat domains to undergo such a conformational change prior to activation of the channel conductance pathway would account for the highly nonlinear voltage-dependence of channel activation. Support for the involvement of the S4 helix in channel gating has recently been provided by experiments using site-directed mutagenesis to alter progressively the charge on this helix in sodium channels expressed *in vitro* in oocytes (26).

With the exception of the short segment linking repeat domains 3 and 4, the interdomain regions and the amino- and carboxyl-termini of the protein are less highly conserved between sodium channels in different species and between channels in different tissues of the same species (19). The most variable region is that linking domains 1 and 2 (ID1-2). This region varies in length by as much as 185 amino acid residues between sequenced channels; the most striking differences are observed between two forms of sodium channel expressed in rat skeletal muscle. Of interest is the fact that this region appears to be the site of phosphorylation *in vivo* by cAMP-dependent protein kinases (27,28).

Unlike the other interdomain regions, ID3-4 is one of the most highly conserved areas of the primary sequence. This region, which contains a number of conspicuous pairs of positively charged residues, is thought to be involved in the process of channel inactivation. Site-directed mutations in this region give rise to channels that activate and conduct normally but fail to inactivate (26). Likewise, antibodies directed against synthetic oligopeptides corresponding to the sequence of this region also interfere with channel inactivation (29). The primary sequence of ID3-4 is consistent with the known sensitivity of channel inactivation to intra-cellularly applied proteases.


CHANNEL TERTIARY STRUCTURE

Computer analysis of sodium channel primary sequences has given rise to a number of proposed models for channel secondary structure and for protein folding and organization into a functional tertiary structure (17,23,24,30). The most popular of these models project six transmembrane helices in each of the four repeat domains, and configure the four domains themselves in a pseudosymmetrical orientation around a central ion pore (17,24) (figure 2A). The pore itself is formed from the interaction of amphipathic helices contributed by each of the domains, with polar residues of these helices facing into the aqueous interior of the pore and nonpolar residues interacting with other transmembrane helices in their respective domains. The S4 helix in each domain is placed in the middle of the helical bundle and does not form part of the pore lining (figure 2B). This structural model is formally similar to current models of the nicotinic acetylcholine receptor ion channel (AchR), with the internal repeat domains of the sodium channel corresponding to the homologous but separate subunits of the AchR.

Although such models of sodium channel structure are intellectually stimulating, their principal value lies in suggesting specific hypotheses that can be tested experimentally. Validation of the tertiary structure proposed by these models requires the use of a range of biochemical, immunological, and molecular biological approaches. For example, one model of tertiary structure may predict that a particular portion of the primary sequence lies on the extracellular face of the membrane while another predicts an intracellular location. Experiments that can differentiate between these possibilities will help to restrict the universe of acceptable models. Since location of one portion of the sequence often constrains the options for adjacent sections, one definitive experiment can have implications beyond the particular segment being probed.

One approach to the question of tertiary structure is to define such discriminant peptides, generate antibodies to them, and identify the binding site of these antibodies in oriented preparations of the channel. With the availability of the primary sequence, peptides of 15 to 30 amino acids corresponding to the regions of interest can be synthesized and used as antigens to prepare such antisera. Colloidal gold-labeled second antibodies can be used to identify antigen-antibody complexes at the electron microscopic level.

This technique has been used to localize the carboxy-terminus of the sodium channel protein to the cytoplasmic side of the membrane in vivo (31). If the folding pattern within each of the four repeat domains is assumed to be comparable, then the even number of putative transmembrane helices within each leads to the prediction that the interdomain regions and the amino-terminus of the protein should be on the cytoplasmic side of the membrane as well. This proposed folding pattern was further tested by localization of a segment of protein within ID2-3 using the same approach of site-directed antibody binding (32). This region was also shown to be on the cytoplasmic surface of the membrane.

Additional information on channel tertiary structure can be obtained from an analysis of proteolytic cleavage patterns for the protein in its native membrane environment (33) as well as in mixed micellar form after solubilization and purification (34). If a number of different proteases are used, these patterns can localize regions of the protein that are most accessible to these enzymes in aqueous solution and hence on the external surface of the protein. In addition, data on the rate of proteolysis and the size of released fragments can provide insight into the compactness of protein structure in these regions.

Such studies have shown that the carboxy-terminal 10,000 daltons of the sodium channel are particularly susceptible to proteolysis in a finely graded fashion suggesting a loosely organized and accessible structure (34). The amino terminus, on the other hand, appears tightly structured and more resistant to proteolytic cleavage. ID1-2 and ID2-3 show intermediate sensitivity to proteolysis. The domains themselves are particularly resistant, suggesting that their structure is largely intramembrane and sequestered from enzymatic attack, and that the short loops joining the transmembrane segments are buried within the protein structure.

Sodium channels can be phosphorylated by cAMP-dependent protein kinases (PKa's) (28,35). Since this phosphorylation takes place in vivo and is catalyzed by cytoplasmic enzymes, the sites of phosphorylation can be defined as lying on exposed portions of the intracellular face of the protein. Phosphorylation by PKa in rat brain sodium channels occurs on serine residues within the ID1-2 (27). Pka-dependent phosphorylation in the rat muscle channel is also restricted to serine residues and appears to be localized within this region (28). Thus, the ID1-2 loop must be exposed on the cytoplasmic surface of the membrane, a location consistent with the data summarized above.

A coherent picture of sodium channel structure is beginning to emerge. The six transmembrane helical segments within each repeat domain are compactly arranged in the plane of the membrane; the amino- and carboxyl-terminal ends of each domain, as well as the amino- and carboxyl-terminus of the entire protein, are located on the cytoplasmic side of the membrane. Glycosylation is probably confined to the larger loops between the S5 and S6 transmembrane helices, which are on the extracellular side of the membrane. The interdomain regions form cytoplasmic loops with varying degrees of compact organization. No direct confirmatory physical information is yet available on the ion pore itself, but the concept of a central pore formed by contributions from each of the repeat domains seems the strongest.

SODIUM CHANNEL SUBTYPES AND CHANNEL DIVERSITY

Although the similarities between sodium channels in evolutionarily distant species is striking, there is clear evidence for the existence of closely related but structurally or functionally distinct sodium channels within a single species or even within a single tissue (1). In mammalian skeletal muscle, for example, sodium channels are blocked by u-conotoxin while brain sodium channels with virtually identical electrophysiological characteristics are insensitive to this toxin (36). Channels in innervated adult skeletal muscle are blocked by nanomolar concentrations of TTX while sodium channels in the same muscle several days after denervation are resistant to TTX at micromolar concentrations (37,38). Sodium channels in normal cardiac muscle are also resistant to TTX. Thus, there must be regions of the protein structure that are important in defining the unique properties of these closely related channel subtypes.

These subtle differences between sodium channel isoforms could arise at a post-translational level, but at least some reflect information encoded in the genome. Three complete sodium channel messages have been cloned and sequenced in rat brain (17); two additional sodium channel sequences have been determined in rat skeletal muscle (18,19). One of these appears identical to the sequence of another sodium channel obtained from rat cardiac muscle (20). The distribution of substitutions among these sequences clearly points to their transcription from separate genes rather than through differential splicing of exons within a single gene. Thus, sodium channels exist as a multigene family in the rat.

A slightly different situation prevails in *Drosophila* where at least two distinct genes encoding sodium channel-like proteins are found (22,39). One of these, the *para* locus, can give rise to several related channel transcripts by differential splicing of exons. This property is also seen with another *Drosophila* ion channel gene encoding the A-current $K^+$ channel.

When the various sequences for sodium channels from the rat, from eel, and from *Drosophila* are compared, regions in their sequence with unusually wide divergence can be identified (Table 1). As noted above, the four repeat domains and ID3-4 exhibit the highest sequence conservation. The amino- and carboxyl-termini show less conservation and generally become more divergent as the ends of the molecule are approached. The most variability is seen in ID1-2. Here the channels fall into two broad

TABLE 1

HOMOLOGIES BETWEEN AMINO ACID SEQUENCE OF SkM2
AND OTHER SODIUM CHANNELS (Per Cent)[1]

| | N | D-1 | ID 1-2 | D-2 | ID 2-3 | D-3 | ID 3-4 | D-4 | C |
|---|---|---|---|---|---|---|---|---|---|
| | | | | | SkM2 | | | | |
| SkM1 (18) | 58 | 72 | 14 | 80 | 23 | 80 | 83 | 74 | 51 |
| RBSCI (17) | 53 | 75 | 44 | 78 | 27 | 77 | 91 | 77 | 53 |
| RBSCII (17) | 53 | 74 | 40 | 80 | 27 | 79 | 89 | 77 | 54 |
| RBSCIII (56) | 55 | 76 | 33 | 80 | 26 | 79 | 89 | 75 | 53 |
| Eel (16) | 45 | 61 | 12 | 74 | 16 | 72 | 76 | 71 | 34 |
| D.m. (39) | -- | 42 | -- | 52 | -- | 52 | -- | 51 | -- |

[1] Homology was calculated considering only exact matches and excluding conservative substitutions. Deletions were considered as a single non-identity independent of length. D.m. = *Drosophila melanogaster*.

groups; one with ID1-2 lengths less than 160 amino acids and the other with lengths of more than 275 residues. In addition, even among channels with segments of comparable length, the sequences are poorly conserved. This is especially striking since the ID1-2 region contains multiple sites for covalent modification in some channel subtypes that are completely deleted in other channel subtypes within the same animal. Particularly striking is the comparison of the TTX-sensitive skeletal muscle sodium channel (18), which has the shortest reported ID1-2 interdomain, and the putative TTX-insensitive sodium channel from the same tissue (19) and from heart (20), which have one of the longest.

The ID1-2 cytoplasmic loop may be the primary site for specific modifications that determine the unique characteristics of a given sodium channel subtype. The possibility has also been raised that this loop is the site of association with the mammalian beta subunit(s).

## VOLTAGE-DEPENDENT CALCIUM CHANNELS

Various classes of voltage-dependent calcium channels have been identified on the basis of their physiological and pharmacological properties (for review, see ref 40). One class of calcium channel proteins, thought to be the L-type channel, has been isolated using its affinity for pharmacological modifiers such as the dihydropyridines as identifying labels (41,42). Biochemically, these channels differ from the voltage-dependent sodium channel in having two large subunits of 165 and 170 kDa designated alpha$_1$ and alpha$_2$ (see 43 for review). These two large subunits each have distinct biochemical properties (44,45,46). Three smaller subunits of 55 kDa (beta), 30 kDa (gamma) and 27-29 kDa (delta) are also present (45).

From a biochemical perspective, this protein appears quite different from the sodium channel proteins discussed above. Recently, however, the calcium channel alpha subunit has been cloned and sequenced (47). This sequence contains a single open reading frame encoding a protein with a predicted molecular weight of 210,000, about the same size as an alpha subunit of the sodium channel. When the calcium channel alpha$_1$ subunit sequence is aligned with sodium channel protein sequences, a remarkable degree of homology is found. The calcium channel alpha$_1$ subunit contains four internal repeat domains, each with the same six putative trans-membrane helices as the sodium channel (47). Most striking is the presence in the calcium channel of an amphipathic S4 helix in each domain at the same location as in the sodium channels and containing the same repeat motif of a lysine or arginine residue at every third position. Indeed, the sequence conservation between sodium and calcium channels in this region is particularly high. Once again, the interdomain regions are divergent and they have little homology with the sodium channel. Of particular interest is the absence of significant homology in the ID3-4 region.

Thus, although functionally and kinetically quite different, the calcium and sodium channels are clearly derived from the same primordial structure. The degree of sequence conservation suggests that their point of divergence must be relatively recent in evolutionary terms. Based on an examination of the calcium channel sequence, one would predict that the basic ion channel of this protein, as well as its voltage-dependent gating properties, are contained within the $alpha_1$ subunit; the other subunits may play a role in interaction with cytoskeletal elements, with modulation of channel activity, or with intracellular trafficking during synthesis. This conjecture has recently been confirmed by experiments in which oocytes were injected with mRNA encoding only the $alpha_1$ subunit of the cardiac calcium channel. These oocytes subsequently expressed functional voltage-dependant calcium channels demonstrating that here, as with the sodium channel, the key physiological elements of the channel are contained within a single large polypeptide chain (48).

## VOLTAGE-DEPENDENT POTASSIUM CHANNELS

Additional insight into the structure of voltage-dependent ion channels has been provided by recent studies of the gene locus encoding the A-current potassium channel in *Drosophila*. Defects in this locus produce an abnormal phenotype known as *shaker*. This phenotype is deficient in the skeletal muscle potassium channel responsible for the voltage-dependent A-current. The identification of these mutants has recently led to the cloning of the gene in which the mutation resides, and to the identification of its gene products (49-51).

The *shaker* locus encodes a series of related messages ranging between 5 and 10 kb. These messages are produced by differential splicing of the various exons within the gene locus (52,53). Transcripts corresponding to these individual message forms can be expressed *in vitro* in oocytes, and produce voltage-dependent potassium currents similar to the A-current (54,55). The exact characteristics of kinetics exhibited by each channel reflect the particular transcript used in its expression.

This voltage-dependent potassium channel, at first glance, appears unrelated to either of the preceding two channel proteins. With a gene product size between 64,000 to 74,000 kDa for the variously spliced messages, it contains no large component comparable to the alpha subunits of the sodium or calcium channels. However, examination of the primary sequence of these $K^+$ channel transcripts reveals surprising similarities to the structure of a single repeat domain in either of the two larger proteins (52). Thus, between a hydrophilic amino- and carboxyl-terminal region is a compact hydrophobic region containing six putative transmembrane helices. The fourth helix in this region contains the now-familiar pattern of positively charged residues at every third position. Comparison of the *shaker* gene product sequence with the sodium and calcium channels demonstrates homology centered on the S4 helices and decreasing in extent with distance in either direction. The overall pattern of organization of a single internal repeat domain of the larger channel proteins is recapitulated in the smaller protein component of this K+ channel. The homologies around the S4 helix strongly suggest that all these proteins derived from the same primordial element.

Voltage-dependent K+ channels of the *shaker* A-current type are probably formed from homotetramers or heterotetramers of the *shaker* locus gene products. Once organized in the membrane, these proteins would be structurally analogous to both the sodium and calcium channels, with pseudosymmetric orientation of the individual subunits (domains) around the central ion pore, and the primary role of the S4 helix in voltage-dependent gating.

## SUMMARY

Voltage-dependent ion channels appear to form a large family whose individual structures suggest a lineage to a common ancestral channel protein. These channels share the feature of having a number (usually four or five) of homologous subunits or homologous internal repeat domains. Each contains a positively charged amphipathic helix at a conserved location within each of these subunits or repeat domains; voltage-dependent gating may be a property of this conserved S4 helix. In each channel, the ion pore itself is thought to be formed by contributions from helices of each of the subunits or domains which are themselves disposed in a pseudosymmetrical fashion around the aqueous pore.

In spite of these similarities, each channel type, as well as subtypes within each type, exhibit unique kinetic and pharmacological properties, and varying patterns of tissue and cellular expression. Presumably these unique aspects of channel function are contributed by the variable regions of the protein structure and in this regard the cytoplasmic loops connecting the repeat domains or the amino- and carboxy-termini of the individual subunits may play a particular role.

An appreciation of both the common aspects of channel structure and the unique characteristics of the specific channel of interest will be useful in determining its contribution to the pathobiology of hypertension and in planning therapeutic approaches in which the channel may play a role.

## REFERENCES

1.  Barchi RL. *TINS* 10: 221, 1987.
2.  Hodgkin AL, Huxley AF. *J Physiol* 116: 449, 1952.
3.  Bezanilla F. *J Membr Biol* 88: 97, 1985.
4.  Catterall WA. *Ann Rev Pharmacol Toxicol* 20: 15, 1980.
5.  Ritchie JM, Rogart RB. *Rev Physiol Biochem Pharmacol* 79: 2, 1977.
6.  Barchi RL. *Ann Rev Neurosci* 11: 455, 1988.
7.  Roberts R, Barchi RL. *J Biol Chem* 262: 2298, 1987.
8.  Grishin EV, Kovalenko EV, Pashkov VN, Shamotienko OG. *Membr Biophys USSR* 1: 858, 1984.
9.  Miller JA, Agnew WS, Levinson SR. *Biochemistry* 22:462, 1983.
10. Hartshorne RP, Catterall WA. *J Biol Chem* 259: 1667, 1984.

11. Kraner SD, Tanaka JC, Barchi RL. *J Biol Chem* 260: 6341, 1985.
12. Borsotto M, Norman RI, Fosset M, Lazdunski M. *Eur J Biochem* 142: 449, 1984.
12. Casadei JM, Gordon RD, Barchi RL. *J Biol Chem* 261: 4318, 1986.
13. Tanaka JC, Furman RE, Barchi RL. *In: Ion Channel Reconstitution,* C. Miller (ed). New York: Plenum Publishing Co., pp 277-306, 1986.
14. Hartshorne R, Tamkun M, Montal M. *In: Ion Channel Reconstitution,* C Miller (ed). New York: Plenum Publishing Co., pp 337-362, 1986.
15. Agnew WS, Rosenberg RL, and Tomiko SA. *In: Ion Channel Reconstitution,* C. Miller (ed). New York: Plenum Publishing Co., pp 307-336, 1986.
16. Noda M, Shimizu S, Tanabe T, Takai T, Kayano T, *et al. Nature* 312:121, 1984.
17. Noda M, Ikeda T, Kayano T, Suzuki H, Takeshima H, *et al. Nature* 320: 188, 1986.
18. Trimmer JS, Cooperman SS, Tomiko SA, Zhou J, Crean S, Boyle M, Kallen RG, Sheng Z, Barchi RL, Sigworth FJ, Goodman RH, Agnew WS, Mandel G. *Neuron* 3: 33, 1989.
19. Kallen RG, Sheng Z, Yang J, Rogart R, Chen L, Barchi RL. *Neuron* 4: 233, 1990.
20. Rogart RB, Cribbs LL, Muglia LK, Kaiser MW, Kephart DD. *Proc Natl Acad Sci USA* 86: 8170, 1989.
22. Loughney K, Kreber R, Ganetsky B. *Cell* 58: 1143-1154, 1989.
23. Guy HR and Seetharamulu P. *Proc Natl Acad Sci USA* 83: 508, 1986.
24. Greenblatt RE, Blatt Y, Montal M. *FEBS Lett* 193: 125, 1985
25. Catterall WA. *Science* 242: 50, 1988.
26. Stuhmer W, Conti F, Suzuki H, Wang X, Noda M, Yahagi N, Kubo H, Numa S. *Nature* 339: 597, 1989.
27. Rossi S, Gordon D, Catterall WA. *J Biol Chem* 262: 17530, 1987.
28. Yang J, Barchi RL. *J Neurochem* 54: 954, 1990.
29. Gordon RD, Merrick D, Wollner DA, Catterall WA. *Biochemistry* 27: 7032, 1988b.
30. Kosower EM. *FEBS Lett.* 182: 234, 1985.
31. Gordon RA, Fieles WE, Schotland DL, Angeletti RA, Barchi RL. *Proc Natl Acad Sci USA* 84: 308, 1987.
32. Gordon RA, Li Y, Fieles WE, Schotland DL, Barchi RL. *J Neurosci* 8: 3742, 1988a.
33. Kraner SD, Yang J, Barchi RL. *J Biol Chem* 264: 13273, 1989.
34. Zwerling S, Kraner S, Barchi RL. *Neurosci Abs* , 1989.
35. Costa MC, Casnellie JE, Catterall WA. J Biol Chem 257: 7918, 1982.
36. Cruz LJ, Gray WR, Olivera BM, Zeikus RD, Kert L, Yoshikami D, Moczydlowski E. *J Biol Chem* 260: 9280, 1985.
37. Thesleff S, Vyskogil F, Ward MR. *Acta Physiol Scand* 91: 196, 1974.
38. Rogart RB, Regan LJ. *Brain Res* 329: 314, 1985.
39. Salkoff L, Butler A, Wei A, Scavarda N, Giffen K, Ifune C, Goodman R, Mandel G. *Science* 237: 744, 1987.
40. Hosey MM, Lazdunski M. *J Membr Biol* 104: 81, 1988.
41. Curtis BM, Catterall WA. *Biochemistry* 23: 2113, 1984.
43. Catterall WA. *J Biol Chem* 263:3535, 1988.
44. Hosey MM, Barhanin J, Schmid A, Vandaele S, Ptasienski J, O'Callahan C, Cooper C, Lazdunksi M. *Biochem Biophys Res Commun* 147: 1137, 1987.
45. Takahashi M, Seagar MJ, Jones JF, Reber BFX, Catterall WA. *Proc Natl Acad Sci USA* 84: 5478, 1987.

46. Leung AT, Imagawa T, Campbell KP. *J Biol Chem* 262: 7943, 1987.
47. Tanabe T, Takeshima H, Mikami A, Flockerzi V, Takahashi H, Kangawa K, Kojima M, Matsuo H, Hirose T, Numa S. *Nature (Lond)* 328: 313, 1987.
48. Mikami A, Imoto K, Tanabe T, Niidome T, Mori Y, Takashima H, Narumiya S, Numa S. *Nature* 340: 230, 1989.
49. Papazian DM, Schwarz TL, Tempel BL, Jan YN, Jan LY. *Science* 237: 749, 1987.
50. Baumann A, Grupe A, Ackermann A, Pongs O. *EMBO J* 7: 2457-2463, 1988.
51. Kamb A, Iverson LE, Tanouye MA. *Cell* 50: 405, 1987.
52. Schwarz TL, Tempel BL, Papazian DM, Jan YN, Jan LY. *Nature* 331: 137, 1988.
53. Pongs O, Kecskemethy N, Muller R, Krah-Jentgens I, Baumann A, Kilts HH, Canal I, Llamazares S, Ferrus A. *EMBO J* 7: 1087, 1988.
54. Timpe LC, Schwarz TL, Tempel BL, Papazian DM, Jan YN, Jan LY. *Nature* 331: 143, 1988.
55. Iverson LE, Tanouye MA, Lester HA, Davidson N, Rudy B. *Proc Natl Acad Sci USA* 85: 5723, 1988.
56. Kayano T, Noda M, Flockerzi V, Takahashi H, Numa S. *FEBS Lett* 228: 187, 1988.

# REGULATION OF IONIC CHANNELS BY G PROTEINS

A.M. Brown[†], A. Yatani[†], G. Kirsch[†], A.M.J. VanDongen[†],
B. Schubert[†], J. Codina[‡] and L. Birnbaumer[†‡]

Departments of Molecular Physiology and Biophysics[†]
and Cell Biology[‡], Baylor College of Medicine
One Baylor Plaza, Houston, Texas 77030

Signals may flow through membranes by a pathway composed of three elements: receptor, G protein and effector. At present we know of about 80 different receptors, about 15 different G proteins and about 15 different effectors (1,2). The primary structures of noradrenergic, dopamine, 5-hydroxytryptamine and rhodopsin receptors are similar indicating that the receptors form a family (2). The predicted secondary structures have seven membrane spanning α helices with the greatest differences among receptors occurring in the cytoplasmic linkers (3). The G proteins also form a family in which heterogeneity resides largely in the α subunits for which some 15 different cDNAs have been described (4,2). Two to four forms of the β subunit have been described and there are at least three forms of the γ subunit (4,2). By contrast, the effectors are quite different from each other and there appears to be no homology among adenylyl cyclase (AC) (4), cGMP phosphodiesterase (PDE) and voltage-dependent ionic channels (5,6), although the predicted secondary structures of AC and voltage-gated channels are similar. Within this group, voltage-gated channels clearly are a family (7).

The ionic channels interact with G proteins in either a membrane-delimited, presumably direct, manner like AC or cGMP PDE, or in a cytoplasmic, indirect manner (8). The direct pathway may include a membrane intermediary. Two limiting possibilities are a single intermediary different from the known G protein membrane enzyme target which serves all ionic channels or intermediaries, possibly subunits, unique to each channel. We exclude AC because forskolin (FOR) and cAMP do not produce biphasic effects on $Ca^{2+}$ or $Na^+$ channels whereas isoproterenol (Iso) does (9,10). Nor do FOR and cAMP have any effect on the best-characterized ionic channel G protein target, the atrial muscarinic $K^+[ACh]$ channel (8). Phospholipase C (PLC) may be excluded because it is not activated in the experiments involving $K^+[ACh]$, $Ca^{2+}$, or $Na^+$ channels (2) and phospholipase A2 (PLA2) may be excluded because a functional anti-PLA2 antibody has no effect on muscarinic or purinergic activation of $K^+[ACh]$ channels (12,13).

A single receptor may activate two G proteins and two or more receptors may converge upon a single G protein (Fig. 1) but it has been thought that one G protein regulates only one effector (1). However, recent

*Cellular and Molecular Mechanisms in Hypertension*
Edited by R.H. Cox, Plenum Press, New York, 1991

experiments have shown that a single $G_s\alpha$, where $G_s$ is the stimulatory regulator of AC, can have at least three effectors in one cell; AC, voltage-gated $Ca^{2+}$ and $Na^+$ channels. This has lead us to propose that G proteins organize membrane effectors spatially into networks (9,10,14; Fig. 1). A corollary is that complex temporal organization may occur (9,10). Next we will recall the evidence for a direct effect of a G protein on $K^+[ACh]$ channels which is the prototype for G protein-ionic channel interactions.

### Direct Gating of $K^+$ Channels by G Proteins

In 1958 it was first shown that ACh hyperpolarized the atrial membrane by increasing $K^+$ permeability. A subsequent result of

I.     > 1 Receptor; > 1 G Protein; 1 Effector

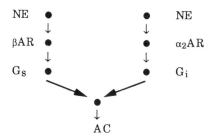

II.     1 Receptor; > 1 G Protein; > 1 Effector

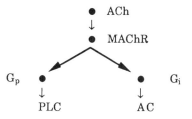

III.     1 Receptor; 1 G Protein; > 1 Effector

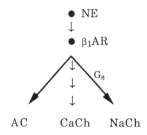

NE, norepinephrine; βAR, β-adrenoceptor; $\alpha_2AR$, $\alpha_2$-adrenergic receptor; $G_s$, the stimulatory G protein regulator of AC; $G_i$, the inhibitory G protein regulator of AC; AC, adenylyl cyclase; MAChR, muscarinic cholinergic receptor; ACh, acetylcholine; $G_p$, G protein regulator of PLC; PLC, phospholipase C; CaCh, a voltage-dependent dihydropyridine-sensitive calcium chan-nel; NaCh, a voltage-dependent sodium channel.

Figure 1: G Protein Membrane Networks.

significance was the latency of 50-150 msec which occurred following topical application of ACh or vagal nerve stimulation (15,16). By contrast, the latency at the neuromuscular junction where the nicotinic AChR (nAChR) and channel are one protein is about 1 msec. The second messengers often considered in such processes are cAMP and cGMP but it was found that neither hyperpolarizes the atrial membrane (17,18).

Binding studies showed that, just as for other receptor-ligand interactions, guanine nucleotides modified the affinity with which muscarinic agonists displace the muscarinic antagonist QNB from mAChR's (19). Furthermore, pertussis toxin (PTX) substrates, sure evidence of the presence of G proteins, were identified in cardiac tissue (20-23). Nevertheless, a connection between guanine nucleotide binding and the latency of the ACh response recorded electrically was not made at this time nor for a considerable time thereafter, and the biochemical and electrophysiological approaches went along separate paths. Then single channel studies identified the inwardly-rectifying single atrial $K^+$ currents that were activated by ACh (24) and the bases for specifying these currents from other single $K^+$ channel currents in atria were established. By perfusing the patch pipette it was shown that ACh activated this $K^+$ channel independently of cytoplasmic mediators (25). This important observation was followed by the report of Pfaffinger *et al* (26) which showed that the ACh response was blocked by PTX and that GTP was required for the ACh effect. At the same time, Breitwieser and Szabo (27) showed that the ACh response became irreversible in the presence of the non-hydrolyzable GTP analog GMP-P(NH)P and they called the putative endogenous G protein $G_k$. These studies linked G proteins with the inwardly-rectifying $K^+$ channel activated by ACh, a linkage that became stronger when GTPγS was shown to activate atrial $K^+$ channels in excised, inside-out membrane patches (28). The GTPγS had an absolute requirement for $Mg^{2+}$ in producing this result (29). Direct involvement of a G protein was strongly implicated but the possibility that the G protein was acting indirectly through, for example, a membrane-associated enzyme such as protein kinase C (PKC) had not been specifically excluded although Kurachi *et al* (30) noted that ATP was not required for the GTPγS effect. Nor did the experiments identify the G protein involved other than that it was a PTX substrate. Therefore, G proteins purified from human erythrocytes (31-33) were applied to ACh-sensitive $K^+$ channels in excised, inside-out membrane patches from mammalian atrial muscle (34,35). The experimental approach is shown in Fig. 2. The basic strategy is to study single channel currents which give less ambiguous results than whole-cell currents and to demonstrate the independence of the G protein effect from second messengers. To do this required using excised, inside-out membrane patches or membrane vesicles incorporated into planar lipid bilayers. Only the exogenous PTX-sensitive G protein called $G_k^*$ reconstituted the mACh effect on $K^+[ACh]$ channels. $G_k$ was preactivated with GTPγS and is denoted as $G_k^*$; it was effective at pM concentrations. The molar ratio for binding is about 1 and GTPγS only has effects within the average 10-20 minute patch lifetime at 10 nM or greater. The cholera toxin (CTX)-sensitive G protein, $G_s$, which activates AC and is the other principal G protein purified from human red blood cells (hRBC) had no effect after preactivation with GTPγS (34,35). The G protein transducin was also ineffective even at nM concentrations. The single channel currents which were identified by their ACh (in this case Carbachol or Carb) responsiveness had a slope conductance of about 40 pS and an average open time of about 1.5 msec (Fig. 3). The results were identical for single channel currents activated by: ACh in cell-attached (C-A) patches, ACh and GTP in the bath solution in excised, inside-out patches, GTP$\gamma$S, $G_k^*$, $\alpha_k^*$, and unactivated $G_k$ plus GTP in the presence of ACh. However, $G_k^*$ and $\alpha_k^*$ effects persist even with washes as long as one hour, whereas the ACh effects cease after GTP is removed. $P_o$, in the absence of activation, is nearly zero and activation is produced by an increase in $P_o$; neither conductance nor open times are affected and the frequency of simultaneous openings for the customary two to three channels in each atrial patch was fitted from binomial expectations. ADP-ribosylation with PTX blocked muscarinic activation and the response was reconstituted by the G protein $G_k$ in the presence of GTP. Hence, endogenous $G_k$ cannot be tightly coupled to either the ACh-sensitive $K^+$ or the muscarinic receptor since ADP-ribosylation with PTX would, in this case, have led to permanent loss of receptor-G protein-$K^+$ channel coupling. As noted, the $\alpha$ subunit of $G_k$ preactivated with GTP$\gamma$S, $G_k^*$, was added at pM concentrations and mimicked the muscarinic response while the $\beta\gamma$ dimer was ineffective at nM concentrations. Moreover, GTP plus $\beta\gamma$ could not reconstitute the response whereas GTP plus $G_k$ under identical conditions could. $\beta\gamma$ subunits on occasion inhibited muscarinic responses. Activated $G_o$ from

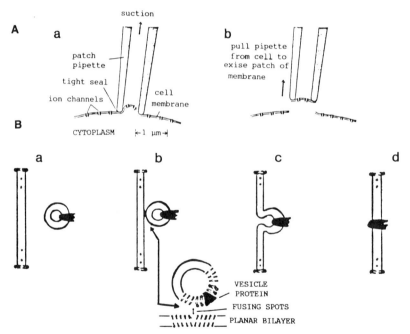

Figure 2: **A)** The patch clamp method. Current enters or leaves the pipette by passing through the channels in the patch of membrane. Recordings of the current through these channels can be made with the patch still attached to the rest of the cell, as in (a), or excised, as in (b). (b) was the method we usually employed. **B)** Method of incorporating membrane vesicles into planar lipid bilayers. The vesicles added to the *cis* chamber are carried to the bilayer by an osmotic gradient (a) and fusion begins at the fusing spots (b-d). The orientation of the $Ca^{2+}$ channels in the vesicle is usually right side out for cardiac sarcolemma and inside out for skeletal muscle T-tubules. The conventions for current recording are the same as those used with patch clamp: positive current is outward.

bovine brain had weak effects which may have been due to contamination with an activated $G_k$ protein (34) but a direct effect of $G_o$ could not be excluded.

The $\alpha$ subunit of the particular PTX-sensitive G protein we tested has not been shown to have an inhibitory effect on adenylyl cyclase. Hence, the G protein was not truly $G_i$. The ACh-sensitive $K^+$ channel was its first proven effector and the active exogenous $G_k$ protein, like the endogenous G protein, was, therefore, called $G_k$ (27), the "k" referring to $K^+$ channel (34). To be effective, the $G_k$ had to be activated either by ACh receptor plus GTP or by GTP$\gamma$S and the first condition required $Mg^{2+}$. Activated $G_k$ ($G_k^*$) was effective in the absence of $Mg^{2+}$ and neither ATP nor AMP-P(NH)P had any effect. Phorbol esters also had no effect. These results excluded phosphorylation via kinases, including PKC, and proved that a specific signal transducing, GTP-activated G protein directly activated an ion channel. The activated $\alpha$ subunit of $G_k$, $\alpha_k^*$, was as effective as $G_k^*$ and the same $\beta\gamma$ subunits were shared by $G_s^*$ and $G_k^*$ yet $G_s^*$ had no effect. $G_k^*$, $\alpha_k^*$, GTP$\gamma$S, and ACh all produced activation in the same way; not by changing channel open time or conductance, but simply by increasing the frequency of channel openings (34,35). The conclusion is that the $\alpha$ subunit mediates the $G_k^*$ effect.

The experiments to this point dealt mainly with reconstitution of the $K^+$[ACh] response. To probe the functional aspects we used a mAb, mAb 4A, that binds to $\alpha_T$ and $\alpha_k$ and had been shown to block rhodopsin-metarhodopsin changes (36,37). The results showed that mAb 4A blocked

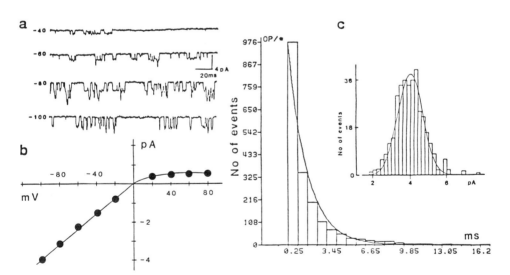

Figure 3: Properties of single muscarinic atrial $K^+$ channel currents obtained from a C-A patch recording with 10 $\mu$M Carbachol in pipette solution. The pipette and bath solutions were isotonic 140 m $K^+$ solution (78). **(a)** representative current traces at indicated holding potentials. Currents were filtered at 1.0 KHz. **(b)** current-voltage relationships for the $K^+$ channel current. Slope conductance was 40 pS. **(c)** open times and amplitude histograms (inset) of single channel currents were obtained at a holding potential of -100 mV.

the mAChR effect and, furthermore, the block was irreversible only if endogenous G protein was activated. Only two conclusions were possible: 1) that $\beta\gamma$ was liberated at a time when the mACh effect was blocked; or 2) that $\alpha GTP\beta\gamma$ and $\beta\gamma$ are equipotent. Neither result is consistent with a role for $\beta\gamma$ in mAChR activation of K$^+$[ACh] (38).

The role of dimeric $\beta\gamma$ remains unresolved. Antibodies that immunoreact with $\alpha^*_i$-3, but not with $\beta\gamma$, block muscarinic activation of K$^+$[ACh] channels (38) and Clapham and Neer (39,40) now agree that the $\alpha$ subunit can mediate the direct G protein effect on ionic channels. Differences with Clapham and Neer (39,40) not with Isenberg (41) persist with regard to the non-specific effects of CHAPS. Clapham and Neer (12) now suggest that $\beta\gamma$ activates PLA$_2$ leading to the production of arachidonic acid and a lipoxygenase metabolite which can stimulate K$^+$[ACh] channels. This pathway is not involved in muscarinic or purinergic responses (12,13). Kurachi (13) finds a requirement for GTP and has a different interpretation of these experiments from Clapham and Neer. Arachidonic acid and its metabolites were used at high concentrations which makes interpretation difficult (42).

As opposed to Clapham and Neer, we have shown that $\beta\gamma$ is inhibitory (43) and is more effective against basally-activated K$^+$[ACh] channels. This has led us to propose that one role for $\beta\gamma$ is to suppress agonist-independent background noise (43). It was also observed in human erythrocytes that none of the $\alpha_i$s nor $\alpha_s$ was preferentially distributed between the two $\beta_{36,35}\gamma$ forms of the dimer. More on the role of $\beta\gamma$ as a mediator of G protein effects on ionic channels will come from the experiments on s-a nodal cell regulation.

While $\alpha^*_k$ appeared to be a single protein in Coomassie blue-stained strips of protein in SDS, when $^{32}$P NAD labelling by ADP-ribosylation with PTX was used, it became clear that $\alpha_k$ was about 95% of the protein in the band and the remainder was another PTX-substrate. Tryptic digest yielded amino-acid sequences predicted by the $\alpha_{i-3}$ cDNA cloned from human liver (44). Therefore, we attempted to express this clone in E. coli cells using the constructs in Mattera et al (45). We found that the protein encoded by $\alpha_{i-3}$, preactivated by GTP$\gamma$S or A1F$^{2-}$, simulated $\alpha^*_k$ effects although at $\approx$ 1-10% of $\alpha^{*'}_k$'s potency. Similar results have been obtained for proteins expressed by $\alpha_s$ cDNA's in the same constructs. In this case, the $\alpha_s$'s also had lower potencies as AC stimulators than did native $\alpha_s$, a result similar to that reported by Graziano et al (46). These results prove that $\alpha_{i-3}$ can act as $\alpha_k$. They do not prove that $\alpha_{i-3}$ is $\alpha_k$ since the effects of the proteins encoded by $\alpha_{i-2}$, $\alpha_{i-1}$, and $\alpha_0$ remain to be tested.

## Direct Gating of Ca$^{2+}$ Channels by G Proteins

K$^+$ channels are not the only channels proven to be regulated directly by G proteins (Table 1). G proteins also directly regulate Ca$^{2+}$ channels from rabbit skeletal muscle T-tubules (47) and from guinea pig or bovine cardiac sarcolemma (48,49). Unlike K$^+$[G$_k$] where G protein gating is obligatory, in the case of Ca$^{2+}$ channels, G protein gating is modulatory; the G protein is not essential for channel opening but membrane depolarization is. Suggestions that G proteins might directly regulate Ca$^{2+}$ channels have been in the literature (50-56). Direct G protein action might have been responsible for these effects, especially since $G_o$ has no known effectors, but $Ca^{2+}$ currents are also reduced by activators of PKC such as OAG and phorbol esters (57,53,52). Thus, although a G protein, possibly $G_o$, mediates the effects, the G protein could be acting indirectly via PKC rather than directly.

To test for direct effects on cardiac $Ca^{2+}$ channels (48,49), GTPγS and G proteins were applied directly just as for the $K^+$ channels. However, the predominant dihydropyridine-sensitive $Ca^{2+}$ channel, variously referred to as the high-threshold, fast-deactivating, long-lasting, or L channel, quickly inactivates or runs down following patch excision. Two approaches were used to deal with this problem. In one approach, ventricular $Ca^{2+}$ channels were activated by Iso or the dihydropyridine agonist Bay K 8644 prior to patch excision. Iso phosphorylates $Ca^{2+}$ channels via PKA (58,59); Bay K 8644 activated the channels directly. Both cause some prolongation of survival time after patch excision, but single channel currents still ceased quickly. In the other approach, $Ca^{2+}$ channels in vesicles from skeletal muscle T-tubules and cardiac sarcolemma were incorporated into planar phospholipid bilayers. Although these channels are DHP-sensitive, the skeletal muscle channels have smaller conductances (19,60,47) and very different kinetics (61,62). Incorporated skeletal muscle T-tubule single $Ca^{2+}$ channel currents do not run down in the presence of Bay K 8644 (63,19) and are stable for relatively long periods when the unincorporated material has been removed (47).

G proteins directly modulate $Ca^{2+}$ channel activity although the effects were clearer in the bilayer experiments (47,49). A single G protein appears to have two effectors and G proteins may, therefore, act as integrators in addition to their role as signal transducers. The $\alpha_s$ subunits are four structures differing by a 15 amino acid insertion/deletion and a serine residue at the carboxy end of the insertion/deletion. The $\alpha_s$ of human erythrocytes is of the short type and it is not known if both forms are present. To determine if one $\alpha_s$ alone could act on both effectors, we expressed the different $\alpha_s$s as recombinants in *E. coli* and tested the purified proteins. We found that each type of $\alpha$ subunit, when preactivated, stimulated both adenylyl cyclase and $Ca^{2+}$ channels proving that one $\alpha_s$ can indeed activate two distinct effectors (45).

## Indirect Gating of Ionic Channels by G Proteins

This category is distinguished from direct G protein gating because, in this case, G proteins have as their effectors membrane enzymes, not ionic channels. The activated enzymes change cytoplasmic constituents and it is via this indirect pathway that ionic channel activity is modulated. Another distinction is that in all instances the effects are modulatory; an obligatory requirement for receptor activation has not yet been shown.

The sequence has been delineated most clearly for rhodopsin-transduction ($G_t$) and voltage-independent $Na^+$ channels in rods and for β-adrenoreceptors-$G_s$ and dihydropyridine-sensitive $Ca^{2+}$ channels in ventricular myocytes. In this context, rhodopsin may be viewed as a receptor and photons as the agonist. Activated rhodopsin promotes the exchange of GDP for GTP on transducin and the released $\alpha_t$ activated a membrane associated cGMP phosphodiesterase. PDE hydrolyses cGMP

**TABLE 1**

**IONIC CHANNELS DIRECTLY GATED BY G PROTEINS**

| Gα Protein | Channel | Receptor | Tissue | References |
|---|---|---|---|---|
| $\alpha_k$, $\alpha_i$-3, r$\alpha_i$-3 | K+, 40 pS, IR | M$_2$, (ACh) | atrium | 11,35,77,78,45 |
| $\alpha_k$, $\alpha_i$-3 | K+, 55 pS, ?R | M$_2$, (ACh), SS | GH$_3$ | 79,80 |
| $\alpha_o$, r$\alpha_o$ | K+, 55 pS, NR | unknown | hippocampus | 81 |
| | K+, 38 pS, NR | unknown | hippocampus | 81 |
| | K+, 38 pS, IR | 5-HT1A, Adenosine | hippocampus | 81 |
| | K+, 13 pS, NR | unknown | hippocampus | 81 |
| $\alpha_i$-1, r$\alpha_i$-1 | K+, 40 pS, IR | M$_2$ (ACh) | atrium | 35,78,45 |
| $\alpha_i$-2, r$\alpha_i$-2 | K+, 40 pS, IRM$_2$ | (ACh) | atrium | 45 |
| $\alpha_s$ | Ca$^{2+}$, DHP-sens., 25 pS | β-AR | atrium, ventricle | 82,49,83,84 |
| $\alpha_s$, r$\alpha_s$, splice variants | Ca$^{2+}$, DHP-sens., 10 pS | β-AR | skeletal muscle | 47,45 |
| $\alpha_s$ | Na+, TTX-sens. | β-AR | atrium, ventricle | 10 |
| $\alpha_i$-3 | Epithelial Na+ | unknown | kidney | 85 |
| $\alpha_i$-3 | Epithelial Cl- | unknown | kidney | 86 |
| $\alpha_i$-3 | K+$_{ATP}$ | unknown | RIN | 87 |
| unknown | K+$_{ATP}$ | unknown | skeletal muscle | 88 |
| $\alpha_i$S | K+$_{ATP}$ | Purinergic | cardiac muscle | 73 |
| unknown | K+$_{Ca}$, 260 pS | β-AR | myometrium | 74 |
| unknown | Ca$^{2+}$, T-type | unknown | dorsal root ganglion | 75 |

which was keeping the $Na^+$ channels open (64). As a result, the channels close and the membrane hyperpolarized (65,66). In rods, the rhodopsin-transducing cGMP PDE are in the disc membranes and the fall in cGMP has to be translated to the plasma membrane where the $Na^+$ channels are located. An analogous sequence is thought to occur in cones, although rhodopsin-$G_t$ and the channel are all in the plasma membrane (67).

In heart, norepinephrine is the agonist for the β-adrenoreceptor and the activated G protein is $G_s$. $G_s$ activates AC, cAMP is increased, and PKA is activated (58,59). The $Ca^{2+}$ channel is presumably phosphorylated just as the DHP receptor in skeletal muscle is (68) and the opening probability is increased. The effects can be prevented by PKI and muscle-specific phosphatase are additional evidence for a phosphorylating mechanism (59,17). This sequence is not involved in basal regulation, however (ibid.).

There are many other examples of indirect G protein gating of ionic channels. The details are not as complete as in the preceding cases, but a similar pattern is present; the differences being the particular G protein involved and the intervening cytoplasmic steps. A significant number involve G protein activation of phospholipase C. DAG and $IP_3$ are formed; DAG activates PKC which may phosphorylate $K^+$ or $Ca^{2+}$ (69) channels, $IP_3$ releases $Ca^{2+}$ from intracellular cells, and the $Ca^{2+}$ can, in turn, alter the activity of a number of membrane channels, in particular $Ca^{2+}$-activated $K^+$ and non-selective cation channels (70,71).

## Spatio-Temporal Regulation of Voltage-Gated Ion Channels

Recently, we have shown the presence of dual cytoplasmic and membrane-delimited $G_s\alpha$ pathways to cardiac $Ca^{2+}$ channels (9,10,14). These were demonstrated by a biphasic (fast, membrane and slow, cytoplasmic) response to $β_1$-adrenergic agonist. Only slow monophasic effects occurred with stimulation downstream from $G_s$, using forskolin, cAMP or IBMX. The response was mediated by $β_1$ adrenoreceptors so the branch point was $G_s$ presumably with $G_s\alpha$ as the mediator.

These experiments lead us to another series of experiments involving the cardiac $Na^+$ channel (10,14). Previous studies have shown that 1) the β-adrenergic agonist Iso decreases maximum upstroke velocity in depolarized ventricular myocytes (72); 2) cAMP-dependent phosphorylation reduces neurotoxin-activated $^{22}Na^+$ flux (27) and promotes inactivation in embryonic rat brain cells (27); and 3) cAMP modulates $Na^+$ currents in frog node of Ranvier (27). These studies suggest that a signal-transducing G protein may link β-adrenergic receptors to $Na^+$ channels. To test this possibility, we examined the effects of Iso on $Na^+$ currents by whole-cell and single-channel recording in neonatal ventricular myocytes from rat. The voltage dependence of the Iso effect involves a hyperpolarizing shift of inactivation that resembles the shift that occurs in response to local anesthetic (73) and, like local anesthetic block, the effect is strongly potentiated by depolarized holding potentials (74). Agents that inhibit inactivation such as pronase, α-scorpion toxin, and a site-directed antibody (75) also act in a voltage-dependent manner but, unlike the G protein effect, they are less effective when the membrane is held at more positive potentials (Fig. 4).

We concluded that: 1) voltage-gated cardiac $Na^+$ channels are directly and indirectly modulated by $G_s$; 2) a single G protein may have as many as

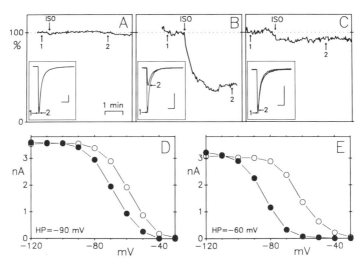

Figure 4: Isoproterenol reduces cardiac $Na^+$ currents at depolarized membrane potentials in a GTP-dependent manner. Primary cardiac cell cultures were prepared from hearts of 1-3 day old neonatal rats. Hearts were removed under sterile conditions and the ventricles were cut into small pieces. The tissue pieces were incubated at 37°C for 5 min in $Ca^{2+}$-free Hanks' solution containing 0.5% trypsin (Sigma T-0134). The supernatant was removed and added to culture medium, DMEM-10 FCS to stop enzyme action. The cell suspensions were seeded on glass cover-slips in 35 mM Falcon dishes containing the culture medium. The cultures were incubated at 37°C in an $H_2O$-saturated, 5% $CO_2$-95% $O_2$ air atmosphere. Cells were used within 24-48 hrs. Whole-cell and single-channel recordings were obtained using patch-clamp techniques. For whole-cell current clamp, we selected small spherical cells (10 μm in dia.) whose membranes behaved as a simple resistance-capacitance (R-C) circuit with a time constant <100 μs. Test potentials were always on the positive limb of the I-V curve, the currents had none of the features of inadequate space clamp and were similar to currents obtained at half $[Na^+]_o$. Patch pipettes had tip resistances of 2.5 MΩ or less, and the input resistance of the cells was about 1.0 GΩ. Capacitative transient cancellation and series resistance adjustments were made to provide optimum settling and attenuation of the capacitive current transient. Currents were digitized and recorded at 44 kHz on a pulse-code modulated video cassette recorder (PCM VCR) for off-line analysis. Before digitization currents were filtered at 5 kHz (-3 dB) using a 4-pole Bessel filter. The data were then transferred to a MicroVax II computer for further analyses. No corrections for leak currents were made for whole-cell recordings. The single channel records were filtered before analysis using a fourth order finite impulse response filter. The experimental chamber (volume 200-500 μl) was placed on an inverted microscope stage. When necessary, external solutions were superfused at 2 ml/min by gravity. To suppress outward currents, the pipettes were filled with $Cs^+$-rich solution of the following composition (mM): CsOH, 118; aspartic acid, 118; $MgCl_2$, 6.4; ethylene glycol-bis (β-aminoethylether)-N,N,N'-N'-tetraacetic acid (EGTA), 5; ATP, 4.2; $CaCl_2$, 2.7; N-2-hydroxyethylpiperazine-N'-2-ethanesulfonic acid (HEPES), 5; pH 7.3 with CsOH; 290 mOsm adjusted with Cs aspartate. Correction was made for liquid junctions potentials between bath solution and pipette solution, 8 ± 0.5 mV, n=6. External solution contained (mM): NaCl, 137; KCl, 5.4; $MgCl_2$, 1; $CaCl_2$, 2; $CoCl_2$, 5; glucose 10; HEPES, 10; pH, 7.4 adjusted with NaOH; 290 mOsm. $Ca^{2+}$ currents were suppressed by addition of $Co^{2+}$ in the presence of $Mg^{2+}$. All experiments were performed at 20-22°. After breaking the patch the pipette solution was allowed to equilibrate with the cell interior for 5 to 10 min. Depolarizing test pulses to 0 mV were applied for 55 msec at 0.5 Hz. Isoproterenol (40 μM, A-C) was applied extracellularly as indicated by the arrow. In A, B, D and E, GTP at 500 μM was included in the pipette solution. Insets in A, B and C show current traces taken at the times indicated. Calibrations are 3.0 msec and 500 pA. A) Holding potential ($V_H$) was -90 mV. The normalized amplitudes of the whole cell $Na^+$ currents are plotted against time. B) Same as A except $V_H$ was -60 mV. C) $V_H$ was -60 mV as in B but GDPγS at 2 mM replaced GTP in the pipette solution. D) $V_H$ was -90 mV. Test pulses to 0 mV were preceded by 200 ms long prepulses of different amplitudes. The $I_{Na}$ amplitude resulting from the test

three effectors; in this case $G_s$ modulates $Na^+$ channels, $Ca^{2+}$ channels, and adenylyl cyclase. Thus, G proteins, in addition to acting as signal transducers, can link different effectors into membrane networks; and 3) in the heart β-adrenergic agonists produce opposite changes in $I_{Ca}$ and $I_{Na}$, but the effects of $I_{Na}$ require that the membrane potential be depolarized. These effects will be exaggerated in the ischemic myocardium because ischemia causes membrane depolarization through extracellular $K^+$ accumulation and is accompanied by an increased catecholamine concentration (76). Thus, our results may explain the data linking high levels of catecholamine to a greater risk of severe arrhythmias associated with myocardial infarction.

## Conclusions

A single G protein may have several effectors and activation by a specific receptor agonist can produce a coordinated response at the membrane level itself. Ionic channels as targets for G protein membrane messengers provide immediate global changes in membrane resting or action potential which may then quickly alter cellular function. In addition, cytoplasmic messengers are activated indirectly and these produce complex cellular responses having a slower time course. Ionic channels are also targets for these indirect $G_k$ protein effects. We, therefore, view the direct interaction between G proteins and ionic channels as a membrane network that can be switched on and off rapidly but which may, through cytoplasmic messengers, produce slower, longer-lasting effects.

## REFERENCES

1. Birnbaumer L, Codina J, Mattera R, Yatani A, Scherer N, Toro M-J, and Brown AM. Signal transduction by G proteins. *Kidney Intl* 32: S14, 1987.
2. Birnbaumer L, Codina J, Yatani A, Mattera R, Graf R, Olage J, Themmen APN, Liao C-F, Sanford J, Okabe K, Imoto Y, Zhou Z, Abramowitz J, Suki WN, Hamm HE, Iyengar R, Birnbaumer M, and Brown AM. Molecular basis of regulation of ionic channels by G proteins. In: *Recent Progress in Hormone Research*, Vol. 45, New York: Academic Press, p 121, 1989.
3. Lefkowitz RJ and Caron MG. Adrenergic receptors. *J Biol Chem* 263: 9887, 1988.
4. Gilman AG. G proteins: transducers of receptor-generated signals. *Ann Rev Biochem* 56: 615, 1987.
5. Frech GC, VanDongen AMJ, Schuster G, Brown AM, Joho RH. A novel potassium channel with delayed rectifier properties isolated from rat brain by expression cloning. *Nature* 340: 642, 1989.

pulse was plotted against the prepulse potential. The points were fitted by a Boltzmann equation and the parameters of the fitted curve were $V_{0.5} = -59$ mV and the slope factor k = 7.3 mV. After application of 40 µM (filled symbols) Iso, the curve was shifted to the left by 9.6 mV without a change in the slope. E) $V_H$ was -60 mV. Pulse protocol was the same as in D. Open symbols correspond to the curve before and closed symbols correspond to the curve after application of 40 µM Iso. The curve after Iso application was shifted by -23 mV with a decrease in slope factor from 7.7 mV to 6.8 mV.

6. Perez-Reyes E, Kim HS, Lacerda AE, Horne W, Wei X, Rampe D, Campbell KP, Brown AM, Birnbaumer L. Induction of calcium currents by the expression of the $\alpha_1$-subunit of the dihydropyridine receptor from skeletal muscle. *Nature* 340: 233, 1989.

7. Numa S, Noda M. Molecular structure of sodium channels. *Ann New York Acad Sci* 479: 338, 1986.

8. Brown AM, Birnbaumer L. Direct G protein gating of ion channels. *Am J Physiol* 23: H401, 1988.

9. Yatani A, Brown AM. Rapid $\beta$-adrenergic modulation of cardiac calcium channel currents by a fast G protein pathway. *Science* 245: 71, 1989.

10. Schubert B, VanDongen AMJ, Kirsch GE, Brown AM. $\beta$-adrenergic inhibition of cardiac sodium channels by dual G protein pathways. *Science* 245: 516, 1989.

11. Yatani A, Codina J, Brown AM, Birnbaumer L. Direct activation of mammalian atrial muscarinic potassium channels by GTP regulatory protein, Gk. *Science* 235: 207, 1987.

12. Kim D, Lewis DL, Graziadei L, Neer EJ, Bar-Sagi D, Clapham DE. G protein beta gamma-subunits activate the cardiac muscarinic K+ channel via phospholipase A2. *Nature* 337: 557, 1989.

13. Kurachi Y, Itoh H, Sugimoto T, Shimizu T, Miki I, Ui M. Arachidonic acid metabolites as intracellular modulators of the G protein-gated cardiac K+ channel. *Nature* 337: 555, 1989.

14. Brown AM, Birnbaumer L. Ionic channels and their regulation by G protein subunits. *Ann Rev Physiol* 52:197, 1990.

15. Glitsch HG, Pott L. Effects of acetylcholine and parasympathetic nerve stimulation on membrane potential in quiescent guinea pig atria. *J Physiol (Lond)* 279: 655, 1978.

16. Hill-Smith I, Purves RD. Synaptic delay in the heart: an ionophoretic study. *J. Physiol. (Lond)* 279: 31, 1978.

17. Trautwein W, Taniguchi J, Noma A. The effect of intracellular cyclic nucleotides and calcium on the action potential and acetylcholine response of isolated cardiac cells. *Pflügers Arch* 392:307, 1982.

18. Nargeot J, Nerbonne JM, Engels J, Lester HA. Time course of the increase in the myocardial slow inward current after a photochemically generated concentration jump of intracellular cAMP. *Proc Natl Acad Sci USA* 80: 2395, 1983.

19. Rosenberg RL, Hess P, Reeves JP, Smilowitz H, Tsien RW. Calcium channels in planar lipid bilayers: insights into mechanisms of ion permeation and gating. *Science* 231: 1564, 1986.

20. Rosenberger LB, Roeske WR, Yamamura HI. The regulation of muscarinic cholinergic receptors by guanine nucleotides in cardiac tissue. *Eur J Pharmacol* 56: 179, 1979.

21. Hazeki O, Ui M. Modification by islet-activating protein of receptor-medicated regulation of cAMP accumulation in isolated rat heart cells. *J Biol Chem* 256: 2856, 1981.

22. Kurose H, Ui M. Functional uncoupling of muscarinic receptors from adenylate cyclase in rat cardiac membranes by the active component of islet-activating protein, pertussis toxin. *J Cyclic Nucleotide Protein Phosphorylation Res* 9: 305, 1983.

23. Halvorsen SW, Nathanson NM. Ontogenesis of physiological responsiveness and guanine nucleotide sensitivity of cardiac muscarinic receptors during chick embryonic development. *Biochemistry* 23: 5813, 1984.

24. Sakmann B, Noma A, Trautwein W. Acetylcholine activation of single muscarinic K$^+$ channels in isolated pacemaker cells of the mammalian heart. Nature 303: 250, 1983.
25. Soejima M, Noma A. Mode of regulation of the ACh-sensitive K channel by the muscarinic receptor in rabbit atrial cells. *Pflügers Arch* 400: 424, 1984.
26. Pfaffinger PJ, Martin JM, Hunter DD, Nathanson NM, Hille B. GTP-binding proteins couple cardiac muscarinic receptors to a K channel. *Nature* 317: 536, 1985.
27. Breitwieser GE, Szabo G. Uncoupling of cardiac muscarinic and β-adrenergic receptors from ion channels by a guanine nucleotide analogue. *Nature* 317: 538, 1985.
28. Kurachi Y, Nakajima T, Sugimoto T. On the mechanism of activation of muscarinic K$^+$ channels by adenosine in isolated atrial cells: involvement of GTP-binding proteins. *Pflügers Arch* 407: 264, 1986.
29. Kurachi Y, Nakajima T, Sugimoto T. Role of intracellular Mg$^{2+}$ in the activation of muscarinic K$^+$ channel in cardiac atrial cell membrane. *Pflügers Arch* 407: 572, 1986.
30. Kurachi Y, Nakajima T, Sugimoto T. Acetylcholine activation of K$^+$ channels in cell-free membrane of atrial cells. *Am J Physiol* 251: H681, 1986.
31. Codina J, Hildebrandt JD, Iyengar R, Birnbaumer L, Sekura RD, Manclark CR. Pertussis toxin substrate, the putative N$_i$ of adenylyl cyclases, is an α/β heterodimer regulated by guanine nucleotide and magnesium. *Proc Natl Acad Sci USA* 80: 4276, 1983.
32. Codina J, Hildebrandt JD, Sekura RD, M. Birnbaumer, Bryan J, Manclark CR, Iyengar R, Birnbaumer L. N$_s$ and N$_i$, the stimulatory and inhibitory regulatory components of adenylyl cyclases. Purification of the human erythrocyte proteins without the use of activating regulatory ligands. *J Biol Chem* 259: 5871, 1984.
33. Codina J, Hildebrandt JD, Birnbaumer L, Sekura RD. Effects of guanine nucleotides and Mg on human erythrocyte N$_i$ and N$_s$, the regulatory components of adenylyl cyclase. *J Biol Chem* 259: 11408, 1984.
34. Yatani A, Codina J, Brown AM, Birnbaumer L. Direct activation of mammalian atrial muscarinic K channels by a human erythrocyte pertussis toxin-sensitive G protein, G$_k$. *Science* 235: 207, 1987.
35. Codina J, Yatani A, Grenet D, Brown AM, Birnbaumer L. The α subunit of G$_k$ opens atrial potassium channels. *Science* 236: 442, 1987.
36. Hamm HE, Deretic D, Hofmann KP, Schleicher A, Kohl B. Mechanism of action of monoclonal antibodies that block the light-activation of guanyl nucleotide binding protein, transducin. *J Biol Chem* 262: 10831, 1988.
37. Deretic D, Hamm HE. Topographic analysis of antigenic determinations recognized by monoclonal antibodies to the photoreceptor guanyl nucleotide binding protein, transducin. *J Biol Chem* 262: 10839, 1988.
38. Yatani A, Hamm H, Codina J, Mazzoni MR, Birnbaumer L, Brown AM. A monoclonal antibody to the α-subunit of G$_k$ blocks muscarinic activation of atrial K$^+$ channels. *Science* 241: 828, 1988.
39. Logothetis DE, Kim D, Northup JK, Neer EJ, Clapham E. Specificity of action of guanine nucleotide-binding regulatory protein subunits on the cardiac muscarinic K$^+$ channel. *Proc Natl Acad Sci USA* 85: 5814, 1988.

40. Neer EJ, Clapham DC. Role of G protein subunits in transmembrane signalling. *Nature* 333: 129, 1988.

41. Cerbai E, Kloeckner U, Isenberg G. The α-subunit of the GTP binding protein activates muscarinic potassium channels of the atrium. *Science* 240: 1782, 1988.

42. Piomelli D, Volterra A, Dale N, Siegelbaum SA, Kandel ER, Schwartz JH, Belardetti F. Lipoxygenase metabolites of arachidonic acid as second messengers for presynaptic inhibition of Aplysia sensory neurons. *Nature* 328: 38, 1987.

43. Okabe K, Yatani A, Evans T, Codina J, Birnbaumer L, Brown AM. βγ-subunits of G proteins inhibit muscarinic K+ channels in heart. *J Biol Chem*, submitted, 1989.

44. Birnbaumer L, Yatani A, Codina J, Mattera R, Graf R, Liao C-F, Themmen A, Sanford J, Hamm H, Iyenger R, Birnbaumer M, Brown AM. Signal transduction by G proteins: Regulation of ion channels as seen with native and recombinant subunits and multiplicity of intramembrane transduction pathways. In: *The Molecular and Cellular Endocrinology of the Testis*, Cook BA and Sharpe RM (eds), New York: Raven Press, 1988, p 35.

45. Mattera R, Yatani A, Kirsch GE, Graf R, Olate J, Codina J, Brown AM, Birnbaumer L. Recombinant α-3 subunit of CT protein activates $T_k$-gated K+ channels. *J Biol Chem* 264: 465, 1989.

46. Graziano MP, Casey PJ, Gilman AG. Expression of cDNAs for G proteins in *Escherichia coli*, two forms of $G_{s\alpha}$ stimulate adenylate cyclase. *J Biol Chem* 262: 11375, 1987.

47. Yatani A, Imoto Y, Codina J, Hamilton S, Brown AM, Birnbaumer L. The stimulatory G protein of adenylyl cyclase, $G_s$, also stimulates dihydropyridine-sensitive $Ca^{2+}$ channels: evidence for direct regulation independent of phosphorylation. *J Biol Chem* 263: 9887, 1988.

48. Yatani A, Codina J, Imoto Y, Reeves JP, Birnbaumer L, Brown AM. Direct regulation of mammalian cardiac calcium channels by a G protein. *Science* 238: 1288, 1987.

49. Imoto Y, Yatani A, Reeves JP, Codina J, Birnbaumer L, Brown AM. The α-subunit of $G_s$ directly activates cardiac calcium channels in lipid bilayers. *Am J Physiol* 255: H722, 1988.

50. Triggle DJ, Skattebol A, Rampe D, Joclyn A, Gengo P. Chemical pharmacology of $Ca^{2+}$ channel ligands. In: *New Insights into Cell and Membrane Transport Processes*, Poste G and Crooke ST (eds), New York: Plenum Publishing, pp 125, 1986.

51. Galizzi J-P, Fossett M, Lazdunski M. Properties of receptors for the $Ca^{2+}$ channel blocker verapamil in transverse-tubule membranes of skeletal muscle. *Eur J Biochem* 144: 211, 1984.

52. Lewis DL, Weight FF, Luini A. A guanine nucleotide-binding protein mediates the inhibition of voltage-dependent calcium current by somatostatin in a pituitary cell line. *Proc Natl Acad Sci USA* 83: 9035, 1986.

53. Holz GG IV, Rane S.G., Dunlap K. GTP-binding proteins mediate transmitter inhibition of voltage-dependent calcium channels. *Nature* 319: 670, 1986.

54. Scott RH, Dolphin AC. Regulation of calcium currents by a GTP analogue: potentiation of (-)-baclofen-mediated inhibition. *Neurosci Lett* 69: 59, 1986.

55. Dolphin AC, Scott RH. Calcium channel currents and their inhibition by (-)-baclofen in rat sensory neurones: modulation by guanine nucleotides. *J Physiol* 386: 1, 1987.

56.  Hescheler J, Rosenthal W, Trautwein W, Schultz G.  The GTP-binding protein, $G_o$, regulates neuronal calcium channels. *Nature* 325: 445, 1987.

57.  Rane SO, Dunlap K.  Kinase C activator 1,2-oleylacetyl-glycerol attenuates voltage-dependent calcium current in sensory neurones. *Proc Natl Acad Sci USA* 83: 184, 1986.

58.  Kameyama M, Hofmann F, Trautwein W.  On the mechanism of β-adrenergic regulation of the Ca channel in the guinea pig heart. *Pflügers Arch* 405: 285, 1985.

59.  Kameyama M, Hescheler J, Hofmann F, Trautwein W.  Modulation of Ca current during the phosphorylation cycle in the guinea pig heart. *Pflügers Arch* 407: 123, 1986.

60.  Ma J, Coronado R.  Heterogeneity of conductance states in calcium channels of skeletal muscle. *Biophys J* 53: 387, 1988.

61.  Cota G, Stefani E.  A fast-activated inward calcium current in twitch muscle fiber of the frog (Rana Montezume). *J Physiol (Lond)* 370: 151, 1986.

62.  Caffrey J, Brown AM, Schneider MD.  Mitogens and oncogenes can block the induction of specific voltage-gated ion channels. *Science* 236: 570, 1987.

63.  Affolter H, Coronado R.  Agonists Bay-K8644 and CGP-28392 open calcium channels reconstituted from skeletal muscle transverse tubules. *Biophys J* 48: 341, 1985.

64.  Stryer L.  Cyclic GMP cascade of vision. *Ann Rev Neurosci* 9: 87, 1986.

65.  Fesenko EE, Kolesnikov SS, Lyubarsky AL.  Induction by cyclic GMP of cationic conductance in plasma membrane rod outer segment. *Nature* 313: 310, 1985.

66.  Haynes L, Yau KW.  Cyclic GMP-sensitive conductance in outer segment membrane of catfish cones. *Nature* 317: 61, 1985.

67.  Mathews G.  Single channel recordings demonstrate that cGMP opens the light sensitive ion channel of the rod photoreceptor. *Proc Natl Acad Sci USA* 84: 299, 1987.

68.  Flockerzi V, Oeken HJ, Hofmann F, Pelzer D, Cavalie A, Trautwein W.  Purified dihydropyridine-binding site from rabbit skeletal muscle t-tubules is a functional calcium channel. *Nature* 323: 66, 1987.

69.  Dunlap K, Holz GG, Rane SG.  G proteins as regulators of ion channel function. *TINS* 10: 241, 1987.

70.  Berridge MJ, Irvine RF.  Inositol triphosphate, a novel second messenger in cellular signal transduction. *Nature* 312: 315, 1984.

71.  Petersen OH.  Calcium channels. *Nature* 336: 528, 1988.

72.  Soejima M, Noma A.  Mode of regulation of the ACh-sensitive K-channel by the muscarinic receptor in rabbit atrial cells. *Pflügers Arch* 400: 424, 1984.

73.  Kirsch GE, Codina J, Birnbaumer L, Brown AM.  Coupling of ATP-sensitive $K^+$ channels to $A_1$ receptors by G proteins in rat ventricular myocytes. *Am J Physiol,* 259: H820, 1990.

74.  Ramos-Franco J, Toro L, Stefani E.  GTPγS enhances the opening probability of $K_{Ca}$ channels from myometrium incorporated into bilayers. *Biophys J* 55: 536a, 1989.

75.  Scott RH, Dolphin AC, Wooton JF.  Photorelease of GTPγS inhibits T-type calcium channel currents in rat dorsal root ganglion neurons. *Biophys J* 55: 37a, 1986.

76.  DiFrancesco D.  Characterization of single pacemaker channels in cardiac sino-atrial node cells. *Nature* 324: 470, 1986.

77. Kirsch GE, Yatani A, Codina J, Birnbaumer L, Brown AM. The $\alpha$ subunit of $G_k$ activates atrial $K^+$ channels of embryonic chick, neonatal rat and adult guinea pig. *Am J Physiol* 23: H1200, 1988.

78. Yatani A, Mattera R, Codina J, Graf R, Okabe K, Padrell E, Iyengar R, Brown AM, Birnbaumer L. The G protein-gated atrial $K^+$ channel is stimulated by three distinct $G_{i\alpha}$-subunits. *Nature* 336: 680, 1988.

79. Yatani A, Codina J, Sekura RD, Birnbaumer L, Brown AM. Reconstitution of somatostatin and muscarinic receptor mediated stimulation of $K^+$ channels by isolated $G_k$ protein in clonal rat anterior pituitary cell membranes. *Mol Endo* 1: 283, 1987.

80. Codina J, Grenet G, Yatani A, Birnbaumer L, Brown AM. Hormonal regulation of pituitary $GH_3$ cell $K^+$ channels by $G_k$ is mediated by its *alpha* subunit. *FEBS Letts* 216: 104, 1987.

81. VanDongen T, Codina J, Olate J, Mattera R, Joho R, Birnbaumer L, Brown AM. Newly identified brain potassium channels gated by the guanine nucleotide binding protein $G_o$. *Science* 242: 1433, 1988.

82. Yatani A, Codina J, Imoto Y, Reeves JP, Birnbaumer L, Brown AM. A G protein directly regulates mammalian cardiac calcium channels. *Science* 238: 1288, 1987.

83. Trautwein W, Cavalie A, Allen TJA, Shuba YM, Pelzer S, Pelzer D. Direct and indirect regulation of cardiac L-type calcium channels by $\alpha$-adrenoreceptor agonists. In: *Advances in Second Messenger and Phosphoprotein Research,* Nishizuka Y (ed), New York: Raven Press, In Press, 1990.

84. Hesslinger B, McDonald TF, Pelzer D, Shuba Y, Trautwein W. Whole-cell calcium current in guinea-pig ventricular myocytes dialysed with guanine nucleotides. *J Physiol,* In Press, 1990.

85. Light DB, Ausiello D, Stanton BA. Guanine nucleotide-binding protein, $\alpha_i^*$-3 directly activates a cation channel in rat renal inner medullary collecting duct cells. *J Clin Invest,* 84: 352, 1989.

86. Schwiebert EM, Light DB, Stanton BA. A G protein, $G_i$-3, regulates a chloride channel in renal cortical collecting duct cells. *J Gen Physiol* 94: 6a, 1989.

87. Ribalet B, Ciani S, Eddlestone GT. Modulation of ATP-sensitive K channels in RINm5F cells by phosphorylation and G proteins. *Biophys J* 55: 587a, 1989.

88. Parent L, Coronado R. Reconstitution of the ATP-sensitive potassium channel of skeletal muscle. Activation by a G protein-dependent process. *J Gen Physiol* 94: 445, 1989.

# EXCITATION-CONTRACTION COUPLING IN THE HEART

Harry A. Fozzard

Cardiac Electrophysiology Laboratories, Departments of
Medicine and the Pharmacological & Physiological Sciences
and the Committee on Cell Physiology
The University of Chicago
Chicago, IL 60637

## INTRODUCTION

Muscle contraction is the development of force or of motion by interaction of two complex proteins, actin and myosin. Their interaction results in relative translation of the thick (myosin) and thin (actin) filaments. The interaction is a chemical association between the head of the myosin molecule and the actin molecule, and force or motion results from a bending of the head where it joins the backbone of the myosin molecule. The chemical interaction itself is permitted as a consequence of a cascade of chemical events resulting from binding of $Ca^{2+}$ to one of a heterotrimeric troponin complex. That interaction causes a change in a tropomyosin molecule, which releases actin sites for reaction with myosin.

This muscle contraction is normally triggered by electrical excitation of the muscle. The excitation process is the transmembrane action potential. Excitation-contraction coupling is the name given to the processes linking the two events. Specifically, it is the process by which the action potential causes a sufficient increase in cytoplasmic $Ca^{2+}$ to activate the contraction.

In general, the $Ca^{2+}$ that interacts with troponin is derived intracellularly from a store called the sarcoplasmic reticulum (SR). These are membrane-bound compartments, filled with $Ca^{2+}$ by an ATP-dependent pump that transports $Ca^{2+}$ from the cytoplasm into the SR compartment. Large amounts of $Ca^{2+}$ can be stored in the SR because they contain a Ca-binding protein called calsequestrin, which chelates $Ca^{2+}$ in a readily releasable form. Release from the SR appears to be through a channel, which is gated by $Ca^{2+}$ entering the cell via voltage dependent Ca channels.

This sequence of events describes excitation-contraction coupling for an individual contraction. However, cardiac muscle normally functions in some steady state of repetitive contractions. Marked differences can occur in the size of contraction depending on the steady rate of activation (the frequency-force relationship), and a complicated sequence of changes occurs after an alteration in the activation rate (the "staircase"

*Cellular and Molecular Mechanisms in Hypertension*
Edited by R.H. Cox, Plenum Press, New York, 1991

135

phenomenon). Further, the cardiac cell can alter its contraction in response to hormone and neurotransmitter actions. In contrast to skeletal muscle, cardiac muscle contraction is regulated at the single cell level over a large range, presumably through modulation by the excitation-contraction coupling process.

Recent progress in excitation-contraction coupling has addressed two major issues. How does the transmembrane action potential cause release of $Ca^{2+}$ from the SR in an individual contraction? How does the cell control the amount of $Ca^{2+}$ in the SR available for release? This chapter will review some of our basic knowledge of coupling of membrane potential with $Ca^{2+}$ release. In heart cells the membrane Ca channels are opened by depolarization, admitting a small amount of $Ca^{2+}$ into the cell. This "coupling" $Ca^{2+}$ gates the SR Ca channel to release a large amount of "activator" $Ca^{2+}$ for interaction with troponin. Several major questions remain about this process. Is the Ca current sufficient to explain entirely the SR release, or is there a separate voltage dependent factor? Secondly, what are the relative roles of the voltage-dependent Ca channels, Na-Ca countertransport, and the sarcolemmal ATP-dependent Ca pump in determining the cellular (SR) load of $Ca^{2+}$ available for release?

## VOLTAGE DEPENDENCE OF CONTRACTION AND THE Ca CURRENT

The early potassium contracture studies of Niedergerke (1) on cardiac muscle and Hodgkin and Horowicz (2) on skeletal muscle clearly demonstrated that contraction is a graded function of membrane potential. More recently, membrane potential could be controlled by voltage clamp, showing that contraction is controlled by the magnitude and duration of depolarization, and by the frequency of depolarization (3,4). The time and voltage range of this dependence is such that the duration of the cardiac action potential and the height of the plateau phase should be important factors in controlling contraction.

The process underlying the voltage dependence of contraction is the voltage dependence of Ca current. Cardiac cells have two types of Ca channels, usually called L and T (5,6). The magnitude of the L type Ca current correlates well with the magnitude of contraction. However, the amount of $Ca^{2+}$ entering the cell through the Ca current is only a small fraction of the amount needed for full contraction (7,8). Estimates of the fraction provided by the Ca current range from 10% to 30%, but it is probably much smaller.

Since the Ca current does not supply all of the $Ca^{2+}$ needed for a particular contraction, the principal source of $Ca^{2+}$ is the SR. In a beautiful series of experiments Fabiato (9) showed that direct application of $Ca^{2+}$ to cardiac cells stripped of their surface membranes causes a graded release of $Ca^{2+}$ from the intact SR. This phenomenon of Ca-induced $Ca^{2+}$ release from the SR completes the excitation-contraction process. In summary, the action potential depolarizes the surface membrane sufficiently to activate voltage-dependent Ca channels. The Ca current through these channels causes a graded release of $Ca^{2+}$ from the SR, and this $Ca^{2+}$ is then free to diffuse to the troponin to begin the contractile protein interaction sequence.

While this description of the mechanism of excitation-contraction coupling is certainly correct, it leaves unanswered several important questions. With regard to the activation of an individual contraction, these include: 1) Is there any direct effect of membrane voltage on the Ca-release mechanism of the SR? 2) What is the mechanism of $Ca^{2+}$ release by the SR? With regard to regulation of the magnitude of the steady state contraction, where the beat-to-beat activation is constant, what are the relative roles of Ca current, Na-Ca exchange, and Ca pumping on the filling of the SR?

## A POSSIBLE ROLE FOR VOLTAGE IN Ca RELEASE?

The correlation between Ca current and activation of contraction identifies a role for the sarcolemmal Ca channel. In skeletal muscle, electrical activation of a Ca channel protein also appears to be important, but in that tissue the channel is physically coupled to the SR release channel, and actual current through the channel is apparently not required for gating. Nabauer *et al* (10) examined this question for cardiac cells by determining the effect of having other ions than $Ca^{2+}$ carry the current through the Ca channels. They used voltage clamp in a single myocyte combined with optical monitoring of intracellular $Ca^{2+}$ ($Ca_i$). Removal of all divalent ions permits very large currents through the Ca channels carried by $Na^+$, but these currents failed to trigger release of $Ca^{2+}$ from the SR. Similarly, $Ba^{2+}$ as current carrier failed to trigger $Ca^{2+}$ release. These experiments do not support the idea of any direct coupling between the Ca channel and the SR release channel. Rather, they show that Ca current itself is necessary.

A fascinating experiment has been reported by Valdeolmillos *et al* (11), who loaded rat ventricular cells with "caged" $Ca^{2+}$. Optical release of this $Ca^{2+}$ induced a contraction, which was blocked by the SR channel blocker ryanodine, but not by Ca channel blockers. This result strongly supports the mechanism of Ca-induced $Ca^{2+}$ release.

Cannell *et al* (12) have also reported a combination of voltage clamp with measures of $Ca_i$. In their studies the voltage threshold for $Ca_i$ increase was more negative than that for the Ca current. Peak $Ca_i$ was maximal at voltages where Ca current was only half maximal, and further increases in Ca current failed to produce a greater rise in $Ca_i$. With reduction of Ca current at more depolarized potentials where the Ca driving force was reduced, peak $Ca_i$ failed to fall until Ca current was about half maximal. They interpreted this to mean that normally the Ca current is supramaximal for $Ca^{2+}$ release. Repolarization of the membrane potential before the end of the $Ca_i$ transient accelerated the decline of $Ca_i$, suggesting that there is a direct effect of voltage on the SR $Ca^{2+}$ release mechanism. One possible way for voltage to affect $Ca_i$ is entry of $Ca^{2+}$ through Na-Ca exchange (operating in the physiologically reversed mode). They ruled this out by showing that a step to +100 mV, near the Ca channel reversal potential, failed to cause increase in $Ca_i$. At that potential $Ca^{2+}$ entry from Na-Ca exchange would be expected, but it was apparently not enough to cause SR release during the brief depolarization.

Some careful quantitative experiments have been reported by Beuckelmann and Wier (13). They found, in contrast to Cannell *et al* (12), a very good correlation between peak Ca current and $Ca^{2+}$ release by the SR over a large range of voltages. The fit was further improved by correlation with the integral of Ca current, which should be a better indicator of actual $Ca^{2+}$ entry. Steps to very positive potentials for short times failed to show increased $Ca_i$, as would be predicted if direct coupling existed between the

Ca channel and the SR release channel. Furthermore, repolarization from very positive potentials during the time the Ca channels were open resulted in a pulse of Ca current and a rise in $Ca_i$ that is incompatible with the coupled channel idea. They and Cannell *et al* (12) also showed reduction in the $Ca_i$ transient with short voltage clamp steps, where repolarization occurred after $Ca_i$ had risen partially, consistent with Fabiato's (14) experiments with skinned cells. This is a surprize, because one would expect that when $Ca^{2+}$ release from the SR begins, it will raise $Ca_i$ sufficiently that further Ca current is not important for further release. If repolarization occurred after the peak of the $Ca_i$ transient, the decline of $Ca_i$ was accelerated, in contrast to Fabiato's (14) experiments. It is not clear whether or not this result is compatible with the model of Ca-induced $Ca^{2+}$ release of the SR. At this point, we do not know enough about the local concentrations of $Ca^{2+}$ in the region of the SR or about the release channel to interpret these experiments further.

In summary, there is strong evidence against a direct coupling between the surface membrane Ca channel and the Ca-release channel of the SR. However, the Ca-induced $Ca^{2+}$ release mechanism is a complex and inadequately understood process, and some voltage dependence cannot yet be completely excluded.

## THE Ca RELEASE CHANNEL OF THE SARCOPLASMIC RETICULUM

We have long suspected from intact muscle or skinned cell experiments that the rapid release of $Ca^{2+}$ from the cardiac SR was via a special channel in the SR membrane. Recently this channel has been characterized biochemically (15) and physiologically (16-18) in cardiac and skeletal muscle. The high conductance channel is Ca-selective, and its probability of opening is greatly increased by increasing $Ca^{2+}$ from nanomolar to micromolar concentrations. It is also blocked by $Mg^{2+}$ and ryanodine, similar to the Ca-induced $Ca^{2+}$ release in cardiac cells after surface membrane removal. There are significant differences in behavior between the cardiac Ca-release channel and the one found in skeletal muscle, consistent with some of the physiological differences between these two muscles. Since the channel can be purified and reconstituted, it will be possible to determine its primary structure, and to determine its behavior in excitation-contraction coupling. Of great importance is this Ca-induced release mechanism and the properties of its recovery after inactivation.

## REGULATION OF CELL LOADING BY THE VARIOUS TRANSPORT PROCESSES

Since cardiac contraction is caused by entry of $Ca^{2+}$ from the outside, repetitive activations would increment the total $Ca^{2+}$ within the cells. Up to a point, the Ca-pump of the SR could maintain diastolic levels of $Ca_i$ by pumping the excess $Ca^{2+}$ into the SR, and this would increase the amount available for release. For a variety of reasons it appears that the Ca current of a normal action potential is supramaximal for SR $Ca^{2+}$ release, so that greater loading of the SR would produce greater release and stronger contraction. However, if cell loading continued indefinitely, the cells would become grossly overloaded with $Ca^{2+}$. Some mechanism for

$Ca^{2+}$ efflux against its electrochemical gradient must be available, if we are ever to have a steady state of $Ca^{2+}$ loading and contraction. This efflux must balance the influx by the Ca current or any other influx process.

A Na-Ca countertransport system was originally described by Reuter & Seitz (19). This mechanism uses the energy of the Na electrochemical gradient to pump $Ca^{2+}$ out of the cell against its gradient in exchange for entry of $Na^+$. Some characteristics of this system and their possible relation to contraction have been presented by Sheu and Fozzard (20). This Na-Ca countertransport system appears to be the major mechanism for $Ca^{2+}$ efflux to balance $Ca^{2+}$ entry through the Ca current in the steady state. It remains uncertain whether Na-Ca exchange sets the diastolic level of $Ca_i$, because $Ca_i$ activation of the transport has a high $K_m$ relative to the $Ca_i$ of the resting cell. An alternative is the ATP-dependent Ca pump (21), which has a higher affinity for $Ca_i$. In support of a major role for Na-Ca exchange is the impressive sensitivity of cardiac contraction to the Na electrochemical gradient, most clearly seen in the positive inotropic effect of Na-K pump blockade by digitalis (22,23).

In his original proposal for an important role for Na-Ca exchange in heart, Mullins (24) suggested that during the plateau of the cardiac action potential the ionic gradients would favor $Ca^{2+}$ entry via the countertransport system. Recently, Barcenas-Ruiz *et al* (13) have shown clearly that $Ca^{2+}$ entry can occur at these voltages. In complementary experiments, Brill *et al* (25) showed that the effect of prolonged depolarization beyond 300 msec resulted in substantial increase in contraction because of $Ca^{2+}$ loading via Na-Ca exchange.

In summary, steady-state contraction strength and the phenomenon of staircase appear to be strongly influenced by a balance between $Ca^{2+}$ entry via Ca channels and Na-Ca exchange during activation and $Ca^{2+}$ exit via Na-Ca exchange during recovery.

## ARE THERE EXCITATION-CONTRACTION COUPLING CHANGES IN CARDIAC HYPERTROPHY?

Ventricular hypertrophy is associated with a sequence of contractile changes. Contractility is preserved or increased under some conditions early in the hypertrophy process, but it can be severely depressed at advanced stages, corresponding to clinical heart failure. Animal models of hypertrophy have offered the opportunity to study some of the cellular processes related to these contractile changes. However, caution must be used in interpreting the results from models, because of differences between the natural process in humans and the models. In addition, physiological and biochemical studies can be difficult to compare between models and between laboratories because measurements are made at different stages in the sequence of hypertrophic changes. Possible changes relevant to excitation-contraction coupling include: 1) alteration in the action potential, 2) alteration in Ca channels, 3) surface-to-volume ratio changes, 4) changes in the sarcoplasmic reticulum Ca pump, the $Ca^{2+}$-release channels, or in $Ca^{2+}$ storage, and 5) changes in the contractile proteins.

One of the first studies of electrophysiological changes in hypertrophy was reported by Bassett and Gelband (26), who used an acute pulmonary hypertension model of right ventricular hypertrophy. They initially reported two changes that have been confirmed and extended over the last decade, including a transient lengthening of the action potential at the time when contraction was reduced, and secondly, a slowing of activation and relaxation of contraction. A logical candidate for the mechanism of these changes would be alteration in the Ca current.

The study of Ca current in hypertrophied cells is technically complex, since the currents must be referenced to the membrane surface area if valid conclusions are to be drawn. Since cells from hypertrophied hearts are larger, they have more membrane. Kleiman and Hauser (27) reported such careful studies in single cells from a cat right ventricular hypertrophy model. Peak Ca current was not increased after normalization for surface area increase. Instead, they found a decrease in the transient outward (K+) current. However, there was a slowing of the final decay of the Ca current.

In contrast to those results, Keung (28) has reported an increase in Ca current from hypertrophied rat ventricular cells. He also found a slowing of the decay of the Ca current, which could be important in explaining the slow time course of relaxation of contraction. Keung's experiments were done with Cs+ replacement of K+, which should have avoided any problems with overlapping transient outward current changes. The reasons for the different results on Ca currents is not clear. The difference may reside in the selection of cells for voltage clamp study or in the methods for determination of the membrane surface area. Kleinman and Hauser (27) performed some hemodynamic measurements and could conclude that the hearts from which they obtained their cells were in the compensated stage, rather than the failed one.

It is not clear what a change in the Ca current should do in hypertrophied cells. We have long been interested in the fact that hypertrophy changes the amount of surface membrane less than it changes the amount of SR or myofibrils. It is possible that a normal Ca current density is not adequate for release of $Ca^{2+}$ from the increased quantity of SR. Such a mechanism would predict that an experimental increase in extracellular Ca should restore normal function.

If the change in decay of the Ca current is important in the slow relaxation of hypertrophied muscle, then the $Ca_i$ transient should be changed. Some interesting observations have been reported by Gwathmey and Morgan (29) that agree with this prediction. They used aequorin to monitor $Ca_i$ in hypertrophied muscle from a ferret pulmonary banding model. The muscle twitch amplitude was diminished, and it was not corrected by increase in external $Ca^{2+}$ concentration. The peak value of the $Ca_i$ transient was not diminished, but the decline of $Ca_i$ was slowed. These results could explain the slow relaxation of cardiac muscle, but it is still not clear if the primary change is in the Ca current or in the SR response. More direct studies of Ca channel density (e.g., with dihydropyridine binding) and of SR $Ca^{2+}$ uptake and release may answer this question.

It is apparent that in hypertrophy there are a number of changes in various steps involved in excitation-contraction coupling. Changes may also occur in the troponin-tropomyosin-actin-my system, but these are not considered further here.

SUMMARY

There has been dramatic progress in our understanding of normal cardiac excitation-contraction coupling and in control of contraction strength, as the result of the new patch pipette method of voltage clamping of single cells and the new methods for monitoring $Ca_i$. Several abnormalities have been shown to exist in hypertrophied muscle; the action potential is changed and the contraction is slower. A kinetic change appears to exist in the L-type Ca current, associated with a slower decay of $Ca_i$. The next few years should bring a much improved understanding of the molecular and cellular basis for the changes of hypertrophy.

REFERENCES

1.  Niedergerke R. The potassium chloride contracture of the heart and its modification by calcium. *J Physiol (Lond)* 134: 569, 1956.
2.  Hodgkin AL, Horowicz P. Potassium contractures in single muscle fibres. *J Physiol (Lond)* 153: 386, 1960.
3.  Gibbons WR. Cellular control of cardiac contraction. *In: The Heart and Cardiovascular System*, H.A. Fozzard, E. Haber, R.B. Jennings, A.M. Katz, H.E. Morgan (eds). New York: Raven Press, pp 747-778, 1986.
4.  Winegrad S. Membrane control of force generation. *In: The Heart and Cardiovascular System*, H.A. Fozzard, E. Haber, R.B. Jennings, A.M. Katz, H.E. Morgan (eds). New York: Raven Press, pp 703-730, 1986.
5.  Bean B. Two kinds of calcium channels in canine atrial cells. *J Gen Physiol* 86: 1, 1985.
6.  Mitra R, Morad M. Two types of calcium channels in guinea pig ventricular myocytes. *Proc Natl Acad Sci* 83: 5340, 1986.
7.  Fozzard HA. Heart: excitation-contraction coupling. *Ann Rev Physiol* 39: 201, 1977.
8.  Morad M, Cleemann L. Role of $Ca^{2+}$ channel in development of tension in heart muscle. *J Mol Cell Cardiol* 19: 527, 1987.
9.  Fabiato A. Myoplasmic free calcium concentration. *J Gen Physiol* 78: 457, 1981.
10. Nabauer *et al* 1989.
11. Valdeolmillos M, O'Neill SC, Smith GL, Eisner DA. Calcium-induced calcium release activates contraction in intact cardiac cells. *Pflüegers Arch* 413:676, 1989.
12. Cannell MB, Berlin JR, Lederer WJ. Effect of membrane potential changes on the calcium transient in single rat cardiac muscle cells. *Science* 238: 1419, 1987.
13. Barcenas-Ruiz L, Beuckelmann DJ, Wier WG. Sodium-calcium exchange in heart: membrane currents and changes in $[Ca^{2+}]_i$. *Science* 238: 1720, 1987.
14. Fabiato A. Rapid ionic modifications during the aequorin-detected calcium transient in a skinned canine cardiac Purkinje cell. *J Gen Physiol* 85: 189, 1985.
15. Campbell KP, Knodson CM, Imagawa T, Leung AT, Sutko JL, Kahl SD, Raab CR, Madson L. Identification and characterization of the high affinity ryanodine receptor of the junctional sarcoplasmic reticulum $Ca^{2+}$ release channel. *J Biol Chem* 262: 6460, 1987.

16. Smith JS, Coronado R, Meissner G. Single channel measurements of the calcium release channel from skeletal muscle sarcoplasmic reticulum: activation by $Ca^{2+}$, ATP and modulation by $Mg^{2+}$. *J Gen Physiol* 88: 573, 1986.

17. Smith JS, Rousseau E, Meissner G. Calmodulin modulation of single sarcoplasmic reticulum $Ca^{2+}$-release channels from cardiac and skeletal muscle. *Circ Res* 64: 352, 1989.

18. Rousseau E, Smith JS, Henderson JS, Meissner G. Single channel and $Ca^{2+}$ flux measurements of the cardiac sarcoplasmic reticulum calcium channel. *Biophys J* 50: 1009, 1986.

19. Reuter H, Seitz N. The dependence of calcium efflux from cardiac muscle on temperature and external ion composition. *J Physiol (Lond)* 195: 451, 1968.

20. Sheu S-S, Fozzard HA. Transmembrane $Na^+$ and $Ca^{2+}$ electrochemical gradients in cardiac muscle and their relationship to force development. *J Gen Physiol* 80: 325, 1982.

21. Caroni P, Carifoli E. An ATP-dependent $Ca^{2+}$-pumping system in dog heart sarcolemma. *Nature* 283: 765, 1980.

22. Fozzard HA, Wasserstrom JA. Voltage dependence of intracellular sodium and control of contraction. *In: Cardiac Electrophysiology and Arrhythmias.* New York: Grune and Stratton, pp 31-57, 1985.

23. Im W-B, Lee CO. Quantitative relation of twitch and tonic tensions to intracellular $Na^+$ activity in cardiac Purkinje fibers. *Am J Physiol* 247: C478, 1984.

24. Mullins LJ. *Ion Transport in the Heart.* New York: Raven Press, p 95, 1981.

25. Brill DM, Fozzard HA, Makielski JC, Wasserstrom JA. Effect of prolonged depolarizations on twitch tension and intracellular sodium activity in sheep cardiac Purkinje fibres. *J Physiol (Lond)* 384: 355, 1987.

26. Bassett AL, Gelband H. Chronic partial occlusion of the pulmonary artery in cats. *Circ Res* 32: 15, 1973.

27. Kleinman RB, Hauser SR. Calcium currents in normal and hypertrophied isolated feline ventricular myocytes. *Am J Physiol* 255: H1434, 1988.

28. Keung EC. Calcium current is increased in isolated adult myocytes from hypertrophied rat myocardium. *Circ Res* 64: 753, 1989.

29. Gwathmey JK, Morgan JP. Altered calcium handling in experimental pressure-overload hypertrophy in the ferret. *Circ Res* 57: 836, 1985.

# REGULATION OF HUMAN CARDIAC MYOSIN

## HEAVY CHAIN GENE EXPRESSION BY THYROID HORMONE

E. Morkin, J.G. Edwards, R.W. Tsika,
J.J. Bahl, and I.L. Flink

Departments of Internal Medicine, Physiology and
Pharmacology, University Heart Center
Arizona Health Sciences Center, Tucson, AZ 85724

## INTRODUCTION

Rat cardiac myosin heavy chain (MHC) genes are regulated in ventricular myocardium by 3,5,3'-triiodo-L-thyronine ($T_3$), which stimulates expression of the $\alpha$-MHC gene and decreases $\beta$-MHC mRNA production (1,2). The protein products of the cardiac MHC genes combine to produce three heavy chain isoforms, $V_1(\alpha,\alpha)$, $V_2(\alpha,\beta)$, and $V_3(\beta,\beta)$, in order of decreasing electrophoretic mobility and $Ca^{2+}$-ATPase activity (3). The relative proportions of these isoforms may be functionally important because the speed of contraction in both cardiac and skeletal muscles has been shown to be related to myosin ATPase (4). Recently, some forms of familial hypertrophic cardiomyopathy have been reported to be linked directly to either a partial duplication of the cardiac MHC genes or to a missense mutation of the $\beta$-MHC gene (5,6)

The actions of $T_3$ are believed to be mediated through nuclear thyroid hormone receptors (TRs) (7), that have been identified as the products of the $\alpha$ and $\beta$ c-$erbA$ protooncogenes (8,9). Little is known about regulation of the human genes, but conservation of sequences in the 5'-flanking regions between the rat (10,11) and human $\alpha$- and $\beta$-MHC genes (12,13) suggests that the human genes may be regulated similarly. Accordingly, $T_3$-responsiveness and receptor specificity of human $\alpha$- and $\beta$-MHC/chloramphenicol acetyltransferase (CAT) fusion genes have been studied in fetal rat cardiomyocytes and CV-1 cells (14-16).

## RESULTS AND DISCUSSION

Addition of $T_3$ stimulated the expression of human $\alpha$-MHC/CAT fusion constructs in myocyte cultures and inhibited the expression of $\beta$-MHC constructs in a dose-dependent manner with $EC_{50}$ values of about $5 \times 10^{-8}$ M and $1 \times 10^{-9}$ M, respectively (14,15). The time course of $T_3$-induced changes in CAT activity of $\alpha$- and $\beta$-MHC fusion plasmids resembled that

*Cellular and Molecular Mechanisms in Hypertension*
Edited by R.H. Cox, Plenum Press, New York, 1991

described for levels of mRNA production by the endogenous genes in cultured heart cells (17).

Transient assays in the heart cell culture system revealed a complex series of *cis*-regulatory elements in the 5'-flanking sequences of the human MHC genes, including in each case a basal promoter element with canonical TATAA and CAAT sequences as well as upstream positive and negative activating regions. The location of functional elements within the human α-MHC promoter is shown in Fig. 1. Deletion of sequences upstream from position -78 resulted in a marked loss of activity, as might be anticipated, because this construct extends only 10 bp 5' upstream from the CAAT and TATAA sequences. Chimeric plasmids pHαMHC-120 and pHαMHC-159 showed higher levels of activity both in the presence and absence of hormone, suggesting the presence of one or more thyroid hormone response elements (TREs). Clone pHαMHC-199 showed a dramatic reduction in $T_3$-inducible activity, indicating the presence of a negative regulatory element. The rise and fall in $T_3$-inducible activity observed with the proximal constructs was repeated by plasmids containing additional 5' upstream sequences.

These results are consistent with a minimum model for $T_3$ regulation of gene expression proposed for the growth hormone gene (18), which consists of a basal promoter followed by a TRE and nearby upstream elements that act to amplify the activity induced by the hormone. The

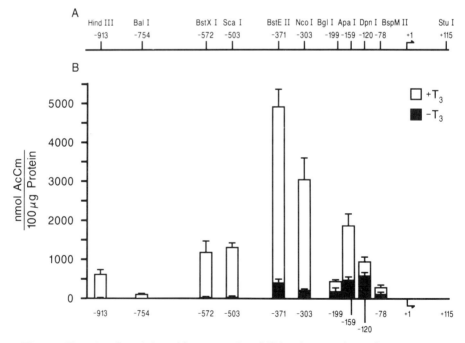

Fig. 1. Functional activity of human α-MHC/CAT fusion plasmids containing deletions in the 5' flanking sequences after transfection into primary cardiomyocytes prepared from 18-19 day fetal rats. All constructs contained a common 3' HindIII site located +115 bp 3' downstream from the transcription initiation site. Cotransfection with pRSVβgal was used as control for transfection efficiency. Values represent mean ± SE for five or six experiments. (from *Proc Natl Acad Sci USA* (14) with permission.)

human α-MHC gene is clearly more complex, however. In addition to one or more TREs, two positive regulatory elements and two negative regulatory elements were identified within the proximal 1 kb of 5' flanking sequence. Interestingly, the positive and negative elements seemed to only regulate $T_3$-induced activity.

Methylation interference and hydroxyl radical footprinting were carried out to more precisely localize the TREs within the α-MHC promoter (19). In the upstream element (TRE$_1$), located at positions -138/-158, the $T_3$ receptors were found to interact within two octameric imperfect direct repeats (underlined) 5'-T<u>CTGGAGGTGA</u>C<u>AGGAGGA</u>CA-3' (anti-sense strand sequence) containing the consensus sequence 5'-C(T/A) GGAGG(T/A)-3'. The sequence of the more proximal $T_3$-binding element (TRE$_2$), located at position -111/-129, is not similar except for a purine-rich octameric cluster (underlined, 5'-ATCAAA<u>GGAGGAGGA</u>GCCA-3') containing six guanines on the sense strand. Binding of liver nuclear $T_3$ receptors to TRE$_1$ was much stronger ($K_d$ = 0.8 nM) than to TRE$_2$ ($K_d$ = 23 nM).

Functional assay of a series of human β-MHC/CAT fusion plasmids also was carried out (Fig. 2). The predominant feature of the β-MHC promoter is an upstream regulatory element that is strongly expressed in the absence of thyroid hormone. Thus the activity of clone pβMHC-293 was more than 10-times greater than that of the adjacent clones (pβMHC-205 and pβMHC-412). Assay of additional constructs have localized the positive promoter element responsible for expression of this gene in the absence of hormone (constitutive expression) between positions -293 and -273 (15). The data also suggests that a suppressor element is present immediately 5' upstream to the strong positive element. It is noteworthy that the activity of pβMHC-293 was not down regulated by addition of $T_3$ to the culture medium whereas both longer and shorter constructs were $T_3$ regulated. The latter results were interpreted to suggest that TREs are located 5' and 3' to the strong positive element.

To further localize the TREs within the proximal 5' flanking sequences of the β-MHC gene, three contiguous unlabeled restriction fragments of 175, 216, and 201 bp, spanning the distance from position -468 to +124, were purified and examined by gel-shift analyses using [$^{125}$I]T$_3$-labeled, partially purified TRs from rat liver. Two TREs were tentatively identified in the human β-MHC gene located on either side of the strong positive element. More recently, by DNase footprinting using bacterially expressed human thyroid hormone receptors, two pairs of TREs have been identified on either side of the strong positive element.

Thus the results of transient assays and footprinting experiments suggest that transcription of the β-MHC gene is driven by a single strong positive promoter element in the proximal 5' flanking sequences. The activity of the positive element is postulated to be regulated by some combination of influences from multiple TREs and the adjacent suppressor element.

The specificity of the TR isoforms for regulation of human cardiac MHC genes by $T_3$ was assessed by cotransfection of α- and β-MHC fusion genes together with expression vectors encoding TR$_{α1}$, TR$_{α2}$, and TR$_{β1}$ into receptor deficient CV-1 cells and fetal rat heart cells (16). Cotransfection with TR$_{α1}$ and TR$_{β1}$ enhanced $T_3$-induced activity of the α-MHC reporter plasmid over the level obtained with endogenous receptors about five-fold in

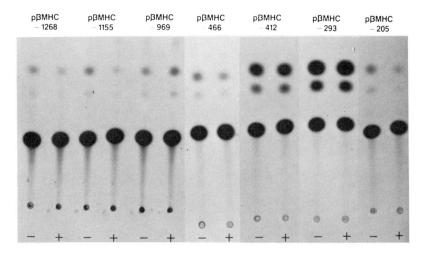

Fig. 2. Activity of human β-MHC/CAT fusion constructs containing deletions in the 5' flanking sequences after tranfection into primary fetal cardiomyocytes. All constructs had a common 3' end located at +124 with respect to the cap site. + and - refers to presence or absence of $10^{-7}$ M $T_3$ in the culture medium.

CV-1 cells and more than 40-fold in cardiomyocytes. The $\alpha_2$ variant, which does not bind $T_3$, reduced $T_3$-induced expression of the α-MHC fusion gene in CV-1 cells, but had no effect in cardiomyocytes. By contrast, cotransfection of β-MHC constructs with $TR_{\alpha 1}$ and $TR_{\beta 1}$ did not change the inhibitory activity of $T_3$ relative to that seen with endogenous receptors in cardiomyocytes. These results suggest that the actions of $T_3$ on the α-MHC reporter gene may be mediated directly through either the $TR_{\alpha 1}$ or $TR_{\beta 1}$ receptor isoforms. Regulation of the β-MHC, however, seems to be more complex and may require additional factors.

## ACKNOWLEDGMENT

The research described in this publication was supported by research grants from the National Institutes of Health (PO1HL20984), the Arizona Affiliate of the American Heart Association/Flinn Foundation, the Arizona Disease Control Research Commission, and the Gustavus and Louise Pfeiffer Research Foundation. J.G.E. was the receipent of an NIH Fellowship (F32HL07603) and R.W.T. was supported by an NIH training grant (HL07249).

## REFERENCES

1.  Everett AW, Sinha AM, Umeda PK, Jakovcic S, Rabinowitz M, Zak R. Regulation of myosin synthesis by thyroid hormone: relative change in the α- and β-myosin heavy chain mRNA levels in rabbit heart. *Biochemistry* 23: 1596, 1984.
2.  Lompre A-M, Nadal-Ginard B, Mahdavi V. Expression of the cardiac ventricular α- and β-myosin heavy chain genes is developmentally and hormonally regulated. *J Biol Chem* 259: 6437, 1984.

3.  Hoh JFY, McGrath PA, Hale PT. Electrophoretic analysis of multiple forms of cardiac myosin: effects of hypophysectomy and thyroxine replacement. *J Mol Cell Cardiol* 10: 1053, 1978.

4.  Swynghedauw B. Developmental and functional adaptation of contractile proteins in cardiac and skeletal muscles. *Physiol Rev* 66: 710, 1986.

5.  Tanigawa G, Jarcho JA, Kass S, Solomon SD, Vosberg H-P, Seidman JG, Seidman CE. A molecular basis for familial hypertrophic cardiomyopathy: An α/β cardiac myosin heavy chain hybrid gene. *Cell* 62: 991, 1990.

6.  Geisterfer-Lowrance AAT, Kass S, Tanigawa G, Vosberg H-P, McKenna W, Seidman CE, Seidman JG. A molecular basis for familial hypertrophic cardiomyopathy: A β cardiac myosin heavy chain missense mutation. *Cell* 62: 999, 1990.

7.  Oppenheimer JH, Schwartz HL, Mariash CN, Kinlaw WB, Wong NCW, Freafe HC. Advances in our understanding of thyroid hormone action at the cellular level. *Endocrine Rev* 8: 288, 1987.

8.  Weinberger C, Thompson CC, Ong ES, Lebo R, Gruol DJ, Evans R. The *c-erb-A* gene encodes a thyroid hormone receptor. *Nature (Lond)* 324: 641, 1986.

9.  Sap J, Munoz A, Damm K, Goldberg Y, Ghysdael J, Leutz A, Beug H, Vennstrom B. The *c-erbA* protein is a high-affinity receptor for thyroid hormone. *Nature (Lond)* 324: 635, 1986.

10. Mahdavi V, Chambers AO, Nadal-Ginard B. Cardiac α and β-myosin heavy chain genes are organized in tandem *Proc Natl Acad Sci USA* 81: 2626, 1984.

11. Edwards JG, Morkin E. Unpublished observation.

12. Saez LJ, Gianola KM, McNally EM, Feghali R, Eddy R, Shows TB, Leinwand LA. Human cardiac myosin genes and their linkage in the genome. *Nucleic Acids Res* 15: 5443, 1987.

13. Yamauchi-Takihara K, Sole MJ, Liew J, Ing D, Liew CC. Characterization of human cardiac myosin heavy chain genes. *Proc Natl Acad Sci USA* 86: 3504, 1989.

14. Tsika RW, Bahl JJ, Leinwand LA, Morkin E. Thyroid hormone regulates expression of a transfected human α-myosin heavy chain gene in fetal rat heart cells. *Proc Natl Acad Sci USA* 87: 379, 1990.

15. Edwards JG, Flink IL, Bahl JJ, Liew CC, Sole MJ, Morkin E. Thyroid hormone regulates the expression of a transfected β-myosin heavy chain gene in rat heart cells. Submitted for publication, 1991.

16. Tsika RW, Edwards JG, Bahl JJ, Morkin E. Regulation of human α- and β-myosin heavy chain genes by thyroid hormone receptor isoforms in receptor deficient CV-1 cells and fetal rat cardiomyocytes. Submitted for publication, 1991.

17. Gustafson TA, Bahl JJ, Markham BE, Roeske WR, Morkin E. Hormonal regulation of myosin heavy chain and α-actin gene expression in cultured fetal rat heart myocyte. *J Biol Chem* 262: 13316, 1987.

18. Glass GK, Franco R, Weinberger C, Alpert VR, Evans RM, Rosenfeld MG. A c-erb-A binding site in rat growth hormone gene mediates *trans*-activation by thyroid hormone. *Nature (London)* 329: 738, 1987.

19. Flink IL, Morkin E. Interaction of thyroid hormone receptors with strong and weak *cis*-acting elements in the human α-myosin heavy chain gene promoter. *J Biol Chem* 265: 11233, 1990.

# REGULATION OF CARDIAC MUSCLE FUNCTION

## IN THE HYPERTENSIVE HEART

Edward G. Lakatta

Laboratory of Cardiovascular Science
Gerontology Research Center, National Institute on Aging
National Institutes of Health
Baltimore, MD 21224

## CARDIOVASCULAR CHANGES IN HUMANS WITH HYPERTENSION

The heart and vasculature of chronically hypertensive patients at any age exhibit cardiovascular changes that occur with aging in individuals whose arterial pressure is within clinically defined normal range. These changes include left ventricular hypertrophy [1-7], a diminution in resting left ventricular early diastolic filling rate [1,2,8,9], increased vascular stiffness [10-16] and aortic impedance [17,18], an increase in peripheral vascular resistance, a diminution in the baroreceptor reflex [19] and a diminished response to catecholamines [20-28].

An increase in arterial impedance, in conjunction with its effect to increase the arterial pulse pressure (and also possibly separately from its effect on arterial pressure [16,29,30]) means that the pulsatile component of external cardiac work must increase. This increase in stroke work appears to be an etiologic factor in the increase in left ventricular wall thickness that occurs in hypertensives. The cardiac hypertrophy is due in large measure to an increase in size of cardiac myocytes, although the connective tissue also increases. In the compensated, adapted hypertensive state the resultant increase in left ventricular thickness tends to normalize myocardial wall stress and permits a normal end-systolic volume and ejection fraction at rest. However, the heart fills more slowly during the early filling period in hypertensives than in normotensives. Prolonged systolic contractile activation (see below) may be a cause of this decrease in the early diastolic filling rate. An enhanced atrial contribution to filling can compensate for the reduction in early diastolic filling and maintains a normal end diastolic filling volume. The augmentation of left atrial filling appears to be related to an increase in left atrial dimension.

### Adaptations in Cardiac Muscle of Experimentally Pressure Loaded Hearts

In conjunction with an increase in myocardial wall thickness and an increase in average cardiac myocyte cell size certain aspects of myocyte function become altered in the hypertensive heart. Although data from the various experimental hypertensive models often cannot be uniformly interpreted because of differences in the age range and/or species investigated, in the mode and duration of pressure overload, or in other aspects of experimental design, some findings common to more than a single model do consistently emerge. These alterations in cardiac muscle from hearts of animals not demonstrating signs or symptoms of congestive heart failure, at rest at least, will be discussed in the context in excitation-contraction coupling mechanisms depicted in Figure 1.

The heartbeat is, in essence, a $Ca^{2+}$ oscillation that is triggered to occur in an organized fashion among myocytes composing the myo-cardium. The rapid increase in cytosolic $[Ca^{2+}]$ $(Ca_i)$ from the resting level (about 100 nM) to systolic levels (from about 0.5 to to 5 μM, depending on the inotropic state) is due to the rapid release of $Ca^{2+}$ from the sarcoplasmic reticulum (SR), triggered by the depolarization induced activation of L type sarcolemmal $Ca^{2+}$ channels during the action potential (Fig 1B). $Ca^{2+}$-myofilament interaction results in myofilament displacement and force production. A second function of SR is to pump the $Ca^{2+}$ it releases into the cytosol back into itself, resulting in a fall in $Ca_i$ to the diastolic level, thus permitting myofilament relaxation. During each heart beat some of the $Ca^{2+}$ released by the SR is lost from the cytosol to the interstitial space via Na/Ca exchange and a sarcolemmal $Ca^{2+}$ pump. In the steady state, a $Ca^{2+}$ influx equal to this loss occurs via $Ca^{2+}$ channels and reverse mode Na/Ca exchange during the action potential.

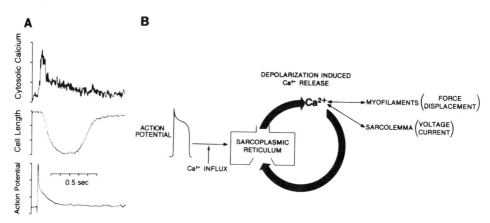

Figure 1: (A) The normal heartbeat at the cellular level. Simultaneous recordings of cytosolic $[Ca^{2+}]$ $(Ca_i)$ measured by Indo-1 fluorescence 410/490 nm (top trace), cell length measured via a photodiode array (middle trace), and transmembrane action potential, measured via a patch pipette (lower trace) in a single isolated rat ventricular myocyte in response to field stimulation (31). (B) Action potential induced $Ca^{2+}$ release from the sarcoplasmic reticulum (SR) occurs via $Ca^{2+}$-induced $Ca^{2+}$ release. $Ca^{2+}$ interacts with myofilament binding sites to cause a contraction and with binding sites within sarcolemmal ionic channels or the Na/Ca exchanger to produce inward current (see text for details). SR is both a source and sink for the $Ca^{2+}$, i.e., it is a "$Ca^{2+}$ oscillator." In addition to triggering the release of $Ca^{2+}$ from SR, the $Ca^{2+}$ current and reverse Na/Ca exchange activated during the action potential place a $Ca^{2+}$ load on the cell and SR. In the steady state, the $Ca^{2+}$ loading is balanced by other sarcolemmal $Ca^{2+}$ extrusion mechanisms (not shown). (From 31).

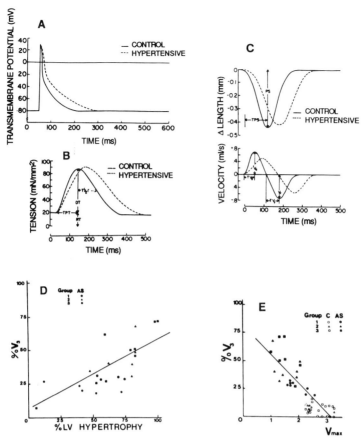

Figure 2: The transmembrane action potential (A), and isometric contraction times (B) are prolonged and isotonic shortening velocity is reduced (C, lower) in left ventricular cardiac muscle isolated from rats with chronic renovascular hypertension compared to that from sham operated rats. (From ref 32 with permission.) An increase in the V₃ form of the myosin heavy chain occurs in pressure loaded rat hearts and varies with the extent of hypertrophy at 8-10 days (O), 1 mo (Δ) and 2 mo (D) and correlates with a reduction in shortening velocity (E). (From ref 40 with permission.)

## ACTION POTENTIAL AND MEMBRANE CURRENTS

The transmembrane action potential is prolonged in all models of compensated pressure hypertrophy in which it has been examined. However, the mechanisms of the prolongation may differ among models. In cardiac muscle from hypertrophied hearts of young *rats with chronic renal hypertension* the transmembrane action potential is markedly prolonged (Fig. 2A) compared with that of muscle from young controls (32-36). While this prolongation is not uniform across the myocardial wall (33,35) it occurs in the absence of changes in electrotonic coupling among cells (36). In single cardiac myocytes isolated from hearts of rats with experimental renovascular hypertension the peak L type $Ca^{2+}$ current ($I_{Ca}$) normalized for cell capacitance surface area (which increases due to the increase in cell size in the hypertrophied heart) is enhanced greater than two-fold (Fig. 3), and the slower phase of current inactivation is prolonged

(28). The steady state voltage inactivation is unaltered. In the *rat aortic constriction model* (37), while the $I_{Ca}$ peak amplitude in single left ventricular cells is two-fold larger, the current ***density*** (peak current normalized for cell capacitance) is unchanged, as are the voltage and time-dependence of $I_{Ca}$ inactivation (38). The number of $Ca^{2+}$ channels in sarcolemmal membranes isolated from hypertrophied and sham-generated hearts as estimated by specific dihydropyridine binding does not differ (39). These data suggest that in general, the ***density*** of $Ca^{2+}$ channels is neither enhanced nor reduced by aortic constriction in rats, implying an enhanced synthesis of the $Ca^{2+}$ channels, commensurate with the increase in cell and heart size (38,39).

Right ventricular pressure overload due to *pulmonary artery constriction* in *cats* is accompanied by an increase in the transmembrane action potential duration measured in intact muscle (41-44) or in myocytes (45). The voltage-dependence and peak amplitude of $I_{Ca}$ measured in single myocytes from hypertrophied hearts did not differ from those in cells from

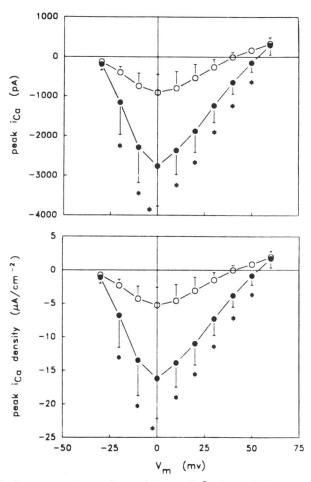

Figure 3: Peak current via sarcolemmal L type $Ca^{2+}$ channels (upper) and current density (lower) are enhanced in single cardiac myocytes from hearts of rats with renovascular hypertension compared to those from sham operated controls. The cells of sham and hypertensive hearts that were studied were equal in size. (From ref 37 by permission of the American Heart Association, Inc.)

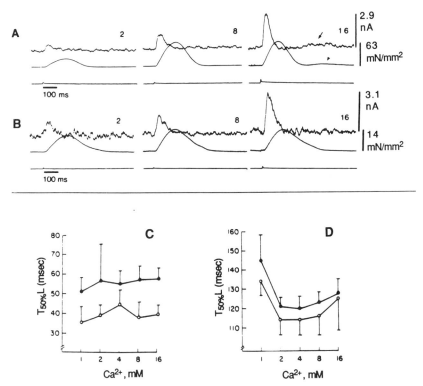

Figure 4: The Ca$_i$ transient, measured by aequorin luminescence, and contraction duration are prolonged in muscle isolated from right ventricle of ferrets with pulmonary artery constriction (A) compared to those from control hearts (B). (From ref 47.) The average prolongation for the Ca$_i$ transient and contraction are depicted in Panels C and D respectively. (From ref 49 with permission.)

control hearts (cells studied from hypertrophied and control hearts were equivalent in size). The integrated $I_{Ca}$ density is slightly increased due to a reduction in the rate of the slower component of current inactivation (45). The cesium-sensitive early outward current was reduced in magnitude (45). Subsequent studies showed that marked alterations in the delayed rectifier K current also occur in this model and include steeper rectification, slower activation and more rapid deactivation. These alterations also likely have a major role in the prolongation of the action potential (46).

The prolonged action potential of hypertrophied cells, regardless of the model, has implications regarding cellular ionic homeostasis and the likelihood for arrhythmias to occur. As Ca2+ influx via L type Ca2+ channels is the physiologic trigger for SR Ca2+ release resulting in the transient increase in Ca$_i$ to cause contraction in cardiac muscle, it may be concluded that in hypertrophied pressure loaded hearts, this trigger is either unaltered or enhanced during the compensated stage, depending on the model studied.

**Cytosolic Calcium Transient**

In ferret cardiac muscle hypertrophied by pulmonary artery banding, the Ca$_i$ transient (Fig. 4), measured as the change in aequorin luminescence following excitation, is not reduced in amplitude but is

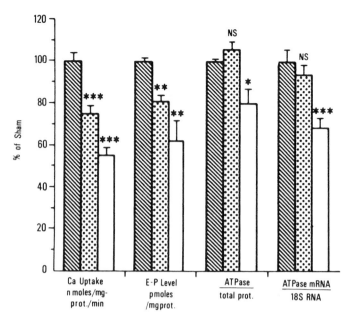

Figure 5: Comparison of the $Ca^{2+}$ transport (nmoles/mg protein/min), phosphorylated enzyme intermediate product (E-P) levels (pmoles/mg protein) and the concentrations of $Ca^{2+}$ ATPase molecules and $Ca^{2+}$ ATPase mRNA in coarctated hearts expressed as a percentage of the same parameters for sham-operated animals. Values are mean ± S.E. Statistical significance of Student's t-test: P versus sham: * < 0.05; ** < 0.001; *** < 0.001; NS, non-significant. Striped columns: sham operated; dotted columns: moderate hypertrophy; empty columns: more marked hypertrophy. (From 51.)

prolonged (48). Preliminary data in isolated myocytes from pressure overloaded cats also indicates a prolongation of the $Ca_i$ transient (48).

A prolonged time course of the $Ca_i$ transient may affect aspects of the cardiac contraction that depend on $Ca^{2+}$-myofilament interactions, including the time to peak force, the relaxation time and the ability of the myofilaments to shorten and stiffen at longer times following excitation. A prolonged $Ca_i$ transient may result from the prolonged action potential or to diminution in the rate of $Ca^{2+}$ removal from the cytosol via the SR and/or Na/Ca exchanger. A prolonged $Ca_i$ transient can also modulate the duration of the action potential. Calcium efflux for $Na^+$ influx occurring during the repolarization phase of the action potential generates an inward current, which prolongs the rat ventricular action potential (50). Exaggerated $Ca^{2+}$ efflux via this mechanism would prolong the action potential of adult rat cardiac muscle, which lacks a delayed K rectifier current, and may be an additional mechanism for the prolonged action potential in rat hypertrophy models.

**Sarcoplasmic Reticulum**

In the rat aortic constriction model the rate of oxalate supported $Ca^{2+}$ uptake and formation of $Ca^{2+}$-dependent phospho-enzyme measured in crude cell homogenates are decreased compared to sham operated animals (Fig. 5). The SR pump protein ($Ca^{2+}$ ATPase) mRNA (measured via a rat cDNA probe) in hypertrophied and control hearts are identical in composition (determined by S-1 nuclease mapping) but the density of the $Ca^{2+}$ ATPase mRNA and of the pump protein itself (measured via specific monoclonal antibody) are reduced in the hypertrophied heart (51). Aortic constriction in rabbits is also associated with a significant reduction in microsomal (SR enriched) $Ca^{2+}$-uptake and $Ca^{2+}$ stimulated ATPase activity (52). Additionally, heat production in cardiac muscle from pulmonary artery banded rabbits in which contraction had been blocked is diminished compared to control muscle, a finding interpreted to reflect a reduction in the rate of SR $Ca^{2+}$ pumping (53). As in the rat aortic constriction model, the SR $Ca^{2+}$ ATPase mRNA in the rabbit pulmonary artery banded model is reduced, to 34% of control (54). Thus, a reduced density of SR pump sites may explain the reduced rate of $Ca^{2+}$ sequestration observed in pressure overload hypertrophy. Phospholamban mRNA was also found to be depressed by the same amount as that for $Ca^{2+}$ ATPase in the rabbit pressure overload model (55).

## Na/K Pump and Na/Ca Exchange

The electrochemical $Na^+$ gradient also modulates the cardiac cell $Ca^{2+}$ load via its influence on the flux through Na/Ca exchanger. Flux through this carrier is dependent on the membrane potential and the sarcolemmal $Ca^{2+}$ and $Na^+$ gradients. The major regulator of the latter is the Na/K pump. In the rat aortic constriction model, an adaptation to pressure overload includes a resurgence of neonatal forms of the Na/K receptor (55,56), This is associated with a 4- to 5-fold decrease in the high and low rates of ouabain binding (55). The relationship of these changes to cell $Na^+$ in the pressure loaded heart is presently unclear, however, as the latter has not been quantified. In the cat pressure overloaded model, the electrogenic activity of the Na/K pump is reduced (57). In the rat aortic stenosis model $Na^+$-dependent $Ca^{2+}$ influx and efflux via the Na/Ca exchanger in isolated sarcolemmal vesicles were found to be markedly depressed (58). A 13-fold reduction in the sensitivity to $Ca^{2+}$ was also measured ($Km = 15$ μM). In itself, this should lead to a marked net reduction of $Ca^{2+}$ efflux via Na/Ca exchange and cause gross $Ca^{2+}$ overload. It is likely that either other adaptations in cell $Ca^{2+}$ homeostasis off set these changes in the Na/Ca exchanger or that the overall conditions in studies of these vesicular preparations do not reflect the *in situ* state.

## Spontaneous Sarcoplasmic Reticulum Calcium Oscillations

As indicated in Figure 1, the normal heart beat is an organized cycling of $Ca^{2+}$ from the SR into the cytosol and back into the SR (Fig. 1A). In a machine of this design, the potential for spontaneous $Ca^{2+}$ recycling or oscillations (CaOs) to occur between organized heartbeats is ever-present. A substantial body of evidence gleaned from a variety of mammalian cardiac preparations indicates that such CaOs can indeed occur (59-69). The probability of CaOs occurrence varies with the $Ca^{2+}$ load available for pumping by SR; thus, regulation of the cell $Ca^{2+}$ content by sarcolemmal ion pumps and carriers is a major determinant of whether CaOs will occur. For example, the presence of catecholamines or inhibition of the Na/K pump (70-73), or situations like acidosis or reflow following myocardial ischemia (67-73), all of which lead to an increase in cell $Ca^{2+}$

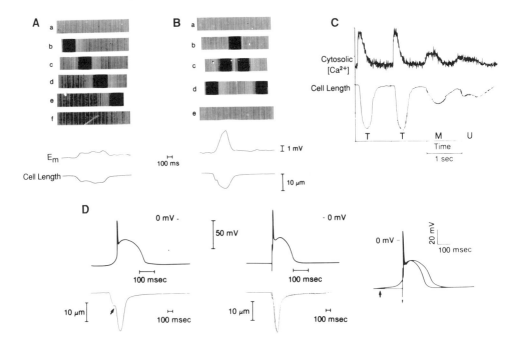

Figure 6: (A, B) Top traces show schematic representation of the localized myofilament shortening, i.e., the contractile manifestation, of localized spontaneous $Ca^{2+}$ release from the SR into the cytosol. The localized myofilament shortening produces a "contractile band" that propagates by diffusion of $Ca^{2+}$, presumably causing $Ca^{2+}$-induced $Ca^{2+}$ release at the wave front, with SR repumping the $Ca^{2+}$ in the wake of the wave (62,63,67). Note that the presence of a contractile band within the cell causes the cell length to decrease from the resting level (see also actual recordings beneath the cartoon). Spontaneous $Ca^{2+}$ release from the SR has been defined as unifocal if, at a given time, it is present in only a single localized area within a cell and thus causes only a single band of contracted sarcomeres (A), or a multifocal if, at a given time, the localized increase in $Ca_i$ is present at more than one locus, giving rise to two or more bands of contracted sarcomeres simultaneously (B). Lower tracings in A and B show actual records of cell length and of membrane potential. Note that the occurrence of the localized increase in $Ca_i$ leads to a depolarization of the sarcolemma. Multifocal SR $Ca^{2+}$ release leads to a greater total area in which $Ca_i$ is elevated and thus a greater extent of myofilament activation of cell shortening and depolarization than the unifocal type. (From 77.) C, Changes in cell length (lower tracing) and in $Ca_i$ (upper tracing) in a rat myocyte during an electrically stimulated twitch (T) and during spontaneous multifocal (M) and unifocal (U) SR $Ca^{2+}$ release following the twitch. The system for measuring $Ca_i$ transients is the same as in Fig. 1, and the signals are unfiltered and unaveraged. The $Ca^{2+}$-dependent fluorescence is measured from the whole cell while the fluorescence underlying U and M is actually instantaneously present in only 15-30 percent of cell area (i.e., the area within the contractile band in A and B). This indicates that the local increase in $Ca_i$ is of the same order of magnitude as that during the stimulated twitch (65). (D) Left Traces, Multifocal spontaneous $Ca^{2+}$ release from the SR (i.e., that similar to that in Panel B) produces a cell shortening (arrow) and causes depolarization sufficient to reach threshold to trigger an action potential, which then causes a twitch contraction. Middle Traces, An electrically stimulated action potential and twitch are shown for comparison. Right Traces, The spontaneous and stimulated action potentials are superimposed to contrast their initiating events (spontaneous depolarization, large arrow; stimulus artifact for stimulated twitch, small arrow). Note the slow changes in membrane potential and cell length owing to spontaneous SR $Ca^{2+}$ release that preceded the action potential and twitch. (From 77.) Montage in Figure 6 adapted from ref. 31.

loading, enhance the probability of CaOs occurrence. The increase in $Ca_i$ caused by CaOs occurrence not only produces diastolic $Ca^{2+}$-myofilament interaction and contraction, but also causes an inward current (Fig. 1) that is due to the $Ca^{2+}$ activation of the sarcolemmal nonspecific ionic channel or of the Na/Ca exchanger (51,74,75). This produces sarcolemmal depolarization (76). Thus, in contrast to the normal heartbeat, in which a SR $Ca^{2+}$ release is triggered by events that result from a depolarization, CaOs *cause* a depolarization.

Spontaneous SR $Ca^{2+}$ release, unlike that triggered by an action potential, occurs locally within cardiac cells and asynchronously among cells. At a given instance, a single local occurrence (Fig. 6A) or multiple loci (Fig. 6B) can be present within a cardiac cell or among cells comprising the myocardium. The local $Ca_i$ achieved can be as high as that triggered by an action potential during systole (Fig. 6C). When CaOs occur simultaneously in multiple foci, a larger depolarization is produced in an individual cell than CaOs is localized within a single area (Panel A and B, lower traces). When this depolarization is of sufficient magnitude (to initiate the opening on $Na^+$ or $Ca^{2+}$ channels, depending on the membrane potential), it can behave as a pacemaker to trigger a spontaneous action potential (Panel D). In an ensemble of cells that constitutes myocardial tissue, the synchronous occurrence of CaOs among a sufficient number of cells is required to produce a depolarization of sufficient magnitude to trigger an extrasystole. This is a mechanism that likely underlies the transient depolarization due to Na-K inhibition, i.e., due to glycoside toxicity (51,74-76). As noted above, the probability for this "partial synchronization" of CaOs to occur depends on the frequency of the CaOs within individual cells, which depends on the extent to which individual cells are $Ca^{2+}$ loaded.

Cardiac muscle of pressure loaded hearts demonstrates the electrical manifestation of spontaneous SR $Ca^{2+}$ release, i.e. diastolic after-depolarizations (DADs), under conditions in which normal myocardium doe not (Fig. 7A) (78,79). Combined alterations in $I_{Ca}$, Na/Ca exchanger, Na-K pump and SR $Ca^{2+}$ pump, possibly in conjunction with nonspecific changes in membrane ionic permeabilities, may predispose to altered cell $Ca^{2+}$ homeostasis that appears to underlie this abnormality in the hypertrophied heart. The greater likelihood for spontaneous CaOs to occur in pressure hypertrophied hearts may render them more likely to exhibit arrhythmias under conditions when cell $Ca^{2+}$ homeostasis is perturbed, e.g. in response to inotropic drugs or during reflow following ischemia. A predisposition to pacing induced ventricular arrhythmias in aortic banded cats has been reported (80). Hypertrophied hearts due to renal hypertension also show an enhanced likelihood to exhibit voltage oscillations on the action potential plateau (Fig. 7B). In contrast to DADs, these early afterdepolarizations (EADs) are due to the removal of inactivation of $Ca^{2+}$ channels during the long action potential plateau (81) and the resultant cytosolic $Ca^{2+}$ transients are driven by the oscillations in $I_{Ca}$ (82). The greater likelihood for plateau oscillations to occur also likely lead to an enhanced susceptibility of arrhythmias under some circumstances in the pressure hypertrophied heart.

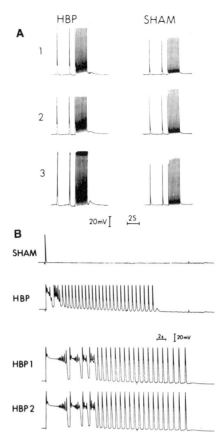

Figure 7: A simultaneous recordings from three sites in muscles from hearts with high blood pressure (HBP) due to renovascular ischemia and sham papillary muscles perfused with 12 mM $[Ca^{2+}]_o$. The last 2 of 10 driven action potentials at a cycle length of 200 msec precede the train of 10 driven action potentials at a cycle length of 200 msec. DADs were recorded at sites 1-3 in the HBP muscle after termination of rapid drive. Although the DADs were synchronous, they varied in magnitude and configuration at each recording site. After depolarizations were not recorded at any site in the sham muscle. (From ref 78.) (B) Triggered activity arising from EAD in HBP papillary muscles in which the action potential was prolonged by blockade of outward currents. Upper tracings, the same external stimulus was applied simultaneously to the sham and HBP muscles following a quiescent period of 2 minutes. The sham preparations responded with only a single driven action potential, whereas the repolarization phase of the driven action potential of the HBP preparation was interrupted by an EAD that gave rise to sustained triggered activity; the triggered activity finally terminates with a DAD. Lower tracings, simultaneous recordings were obtained from 2 sites in the same HBP preparation (interelectrode distance, 1.5 mm). A single driven action potential was evoked after a quiescent period of 2 minutes. The repolarization phase of the driven action potential is interrupted by an EAD which, after an initial quiescent period, gives rise to progressively larger oscillatory responses. The first burst of this triggered oscillatory activity ceases when the membrane repolarizes to a negative level, but then a spontaneous upstroke occurs, apparently from the depolarization. The repolarization phase of this action potential is again interrupted by oscillatory triggered activity. This same sequence is repeated twice more until the membrane potential following repolarization of the fourth burst of oscillatory activity remains a negative level of membrane potential but sustained spontaneous activity follows, presumably having arisen from the depolarizing phase of a DAD. (From ref 78 by permission of the American Heart Association, Inc.)

# FORCE, VELOCITY AND CONTRACTILE PROTEINS

Morphometric investigations of subcellular constituents have shown only small changes in the relative organelle composition of normal and hypertrophied myocytes (83). During the stable, adapted stage of experimentally chronic pressure loading due either to renal hypertension or aortic banding, the isometric force (Fig. 2B) and rate of force development in rodent preparations are preserved, but contraction and relaxation times are prolonged (32,40,84). In contrast to the rat model, chronic pulmonary or aortic banding in ferrets, cats, rabbits or guinea pigs leads to either no change or to a reduction in peak force and to a reduction in the maximum rate of force production as well as to a prolonged contraction in most instances (42,47,85,86). Although the twitch tension is reduced in physiologic bathing medium in muscle isolated from the cats with pulmonary artery banding the maximum tension under conditions that lead to cell $Ca^{2+}$ loading is not altered, suggesting that the diminished twitch tension observed in physiologic milieu is related to an abnormality of excitation-contraction coupling (44). The magnitude of the diminution of twitch tension in physiological bathing milieu was found to correlate with the increase in cell volume which occurred heterogeneously within the myocardium (44). The sensitivity of the force-generating sites within myofilaments for $Ca^{2+}$ can be inferred from the shape of the force-$Ca^{2+}$ relation measured in preparations in which membranous organelles have been destroyed so that the bathing $Ca^{2+}$ concentration can be buffered at constant levels. Over the range of $[Ca^{2+}]$ encountered in the intact cell during contraction, neither the maximum force nor the shape of the force-$Ca^{2+}$ relationship is altered in experimental pressure overloading in rodents, guinea pigs or rabbits (84,87-90).

The isotonic shortening velocity measured at low force loads has been found to decrease in most models of pressure overload regardless of species (Fig. 2C) (32,40,85,91-93) and leads to a shift in the force-velocity relation. A decrease in the rate of ATP hydrolysis has been observed in various contractile protein preparations isolated from the myocardium of pressure-hypertrophied animals (32,84,85,91,94). Myosin ATPase activity is modulated, in part at least, by the myosin heavy chain isoenzyme profile (95), which in pressure-induced hypertrophy in rodents shifts from the predominantly $V_1$ to the predominantly $V_3$ isoform (32,87-90,96-99) (Fig. 2D). This shift is due, in part at least, to a selective increase in the expression of beta-myosin heavy chain mRNA. In ventricles of larger mammals, including humans, $V_3$ is the major isoform (100,101) and no shift occurs during pressure hypertrophy. In human atria, however, the $V_1$ is the major isozyme, and in chronic left atrial hypertrophy associated with valvular disease, a prominent shift to be $V_3$ occurs (101,102). The $V_3$ isoform is energy-sparing in that a given level of tension development in hearts with predominantly the $V_3$ isoform requires less oxygen consumption than in hearts with mixed or predominantly $V_1$ isoforms (85). The reduction in the isotonic shortening velocity in pressure overloaded cardiac muscle is often ascribed to the shift in myosin isoforms (Fig. 2E). However, the slope of the function relating shortening velocity to isoenzyme composition is not unity: in rat preparations in which the % $V_3$ was varied from 0 to 100% only a 40% variation in shortening velocity (103) was shown. Additionally, a substantial increase in cardiac mass and an increase in average cell size by a factor of two resulting from gradual onset pressure overload in rats, which would be expected to markedly shift the myosin to the $V_3$ isoform, did not reduce the velocity of muscle shortening (100). In non-rodent models of pressure overload hypertrophy, the reduction in shortening velocity cannot be ascribed to a myosin isoform shift because, as noted, the $V_3$ form is the predominant form in the ventricular myocardium prior to the hypertrophy and other alterations in myosin, or thin filaments have been evoked (92).

While species/model differences in the extent to which force production or the rate of force production or shortening velocity decreases or the extent to which myosin isoforms shift in the chronically adapted pressure loaded myocardium, relaxation changes are a common feature of all experimental models (32,84,86). Relaxation is regulated by multiple mechanisms, which include the rate of $Ca^{2+}$ dissociation from the myofilaments, the duration of the $Ca_i$ transient, which is dependent on the duration of the sarcolemmal depolarization (action potential) that initiates the increase in $Ca_i$, and the rate of $Ca^{2+}$ pumping into the SR or extrusion from the cell. As noted, in hypertrophied hearts, regardless of species, the action potential is prolonged and the $Ca_i$ transient is prolonged. The $Ca^{2+}$ dissociation rate from troponin C has not been measured. In addition to $Ca^{2+}$-dependent factors, non-$Ca^{2+}$-dependent mechanical factors confer a load-dependence to relaxation and thus can modulate the contraction duration (104). In guinea pig muscle this load sensitivity is markedly reduced and relaxation abnormalities observed have been interpreted to reflect altered $Ca^{2+}$ sequestration (86). These relaxation "abnormalities" can be construed as adaptive in nature as they permit prolonged force bearing capacity required to eject blood into a stiff arterial system, which due to early reflected pulse waves (105) would otherwise tend to abbreviate the ejection time, increase end-systolic volume and reduce ejection fraction.

**Beta-Adrenergic Response**

In the rat aortic constriction model, the inotropic responsiveness to isoproterenol in perfused hearts is depressed (106) (Fig. 8A). Radioligand binding assays (isodopindalol) indicate a decreased beta-receptor density and a normal overall affinity, but a reduction in high affinity sites (106,107). The contractile response to forskolin, which directly activates adenylate cyclase, is also depressed in these hearts (106) (Fig. 8B), suggesting that at least in part, changes in mechanisms distal to the receptor occur in this model. The contractile response to changes in extracellular $[Ca^{2+}]$ is not altered (39). In cardiac cells isolated from aortic constricted rats, the isoproterenol induced increase in $I_{Ca}$ is less than from controls and this difference is abolished by inclusion of cAMP within the patch electrode (38) Deficits in receptor density, adenylate cyclase activation or the myocardial contractile response to beta-adrenergic stimulation have also been observed in the rat Doca-salt and renovascular hypertension models (108-110).

**Extracellular Matrix**

The extracellular matrix is also remodeled in the hypertrophied heart: fibrocytes increase in number and mRNA levels for collagen types and I and III, and exhibit transient increases within 1 to 3 days (111) followed by increases in collagen production which may be heterogeneous in distribution. A peri-vascular fibrosis occurs in the rat hypertensive model which may relate specifically to circulatory renin in this model rather than the hypertension *per se* (112). These changes within the

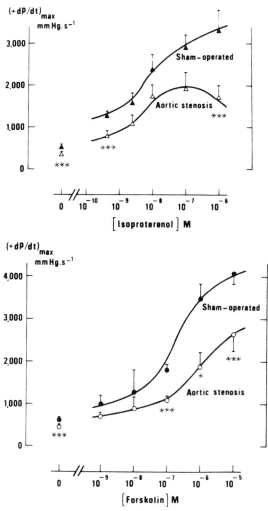

Figure 8: Isoproterenol (A) and forskolin (B) dose-effect curves in sham-operated rats versus those with experimental aortic stenosis. Each point represents the mean ± SEM. *** p < 0.001. N = 3-13. Experiments were performed at a constant coronary pressure of 75 mmHg. (From ref 106 with permission.)

extracellular matrix may alter the mechanical properties, e.g., stiffness, of hypertrophied tissue and thus impact on force production as well as filling properties of the heart.

## CARDIAC MUSCLE CHANGES WITH AGING MIMIC THOSE FOLLOWING EXPERIMENTAL PRESSURE LOADING

With advancing adult age in rodents, ventricular myocytes enlarge (113) and changes in many aspects of cardiac muscle excitation-contraction coupling mechanisms occur (Fig. 9) and are strikingly similar to those in the rodent hypertensive hearts. Specifically, in an isometric contraction the transmembrane action potential (Panel A) (114,115), the $Ca_i$ transient (116), measured as the transient increase in aequorin luminescence (Panel C, inset 1 = aequorin light, 2 = contractile force) and the resultant contraction (Panel B) (32,77,114-125) are longer in duration in cardiac muscle isolated from senescent versus younger adult rats. In individual cardiac myocytes isolated from senescent hearts, the L type $I_{Ca}$ density is not markedly altered but a reduction in its inactivation rate has been observed and may contribute to the prolonged action potential (126). The rate at which the SR pumps $Ca^{2+}$ decreases with aging (Panel D) (121,127) and may be related to a reduction in SR $Ca^{2+}$ ATPase mRNA (128), possibly

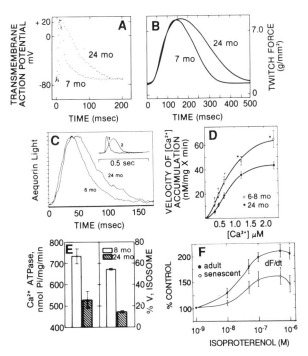

Figure 9: Representative data depicting differences in various aspects of excitation-contraction coupling mechanisms measured between young adult (6 and 9 mo) and senescent (24 to 26 mo) rat hearts. A, Transmembrane action potential; B, isometric contraction; C, $Ca_i$ transient; D, SR $Ca^{2+}$ uptake rate; E, $Ca^{2+}$ stimulated ATPase activity, and myosin isoenzyme composition (50 Per cent of the heterodimer, $V_2$, is included in the total percentage of $V_1$) and F, catecholamine responsiveness. (From 128,129.)

leading to a decrease in the relative density of SR pump sites. The myosin isoenzyme composition shifts to predominantly $V_3$ (21,119) and the myosin ATPase decreases with senescence (Panel E) (32,120). In isotonic contractions, a reduction in speed and extent of shortening occurs in muscle from senescent versus that from younger adult rats (114,118). The time required for restitution of the excitation-contraction coupling cycle is prolonged in senescent muscle, as evidenced by the relative inability to produce a contraction in response to premature stimuli (124).

An age-associated diminution in the effect of beta-adrenergic stimulation on the myocardium has been demonstrated most extensively in the rat model. The beta-adrenergic enhancement of the contractile state in senescent cells (130) or muscle (123) or myocardium (128) is diminished as compared with that from younger adult rats (Fig. 9E). Mechanisms distal to the receptor-cyclase system are required to, at least in part, explain this age-associated diminution of myocardial contractile response to norepinephrine because when dibutyryl cAMP is employed as the agonist, the age deficit in enhancement of contractility observed with beta-adrenergic stimulation persists (128). Additional findings in the Wistar rat aging model indicate that neither the myocardial beta-receptor number nor affinity for antagonists (or for isoproterenol) is altered in advanced age, and neither basal levels of cAMP nor the increased level achieved during the peak contractile response differs with age. These results suggest that perhaps the cAMP-dependent activation of cAMP-dependent protein kinase, which mediates phosphorylation of cell proteins and results in alterations in cell organelle function, may be deficient in the senescent heart. Alternatively, changes in one or more steps distal to protein kinase activation are equally plausible explanations for these results, e.g. in the extent of phosphorylation, in the change in ion flux or binding that results from a given change in phosphorylation. Although neither basal nor stimulated levels of protein kinase activity in the same myocardial preparations in which the contractile response was measured were altered with age, a 20% increase in activity of phosphoprotein phosphatase in the senescent hearts has, in fact, been measured (128). Recently, a diminution in the norepinephrine induced in troponin I phosphorylation in senescent versus adult cardiac myocytes has been demonstrated (131). It is tempting to speculate that since the same *pattern* of changes in cell mechanisms occurs in experimental pressure overload and aging (Fig. 10), this pattern is a result of "logic" within the genome that regulate the expression of multiple genes in order for cellular adaptation to occur. This adaptation allows for an energy efficient and prolonged contraction.

**Transduction Mechanisms of Pressure Loading**

The cardiac hypertrophy itself in pressure overload is a manifestation of enhanced net protein synthesis. Cardiac hypertrophy due to enlargement of a relatively constant number of cardiac myocytes reflects global activation of cardiac genes at the translational and post translational levels (111). There is evidence that the signal that transduces the stress of an enhanced pressure load is mechanical, i.e., stretch or tension (132-138). The stimulation of protein synthesis for both systolic and diastolic tension is described by a single linear function (132). The tension effect may be mediated in part by enhanced coronary flow (139) or hormones (133-135,140). Reorganization of intracellular matrix proteins, desmin and tubulin, may also occur in response to stretch and may mediate the stretch response (141). Stretch of extracellular matrix and ventricular and cardiac

## ALTERATIONS OF CARDIAC MUSCLE FUNCTION
## IN AGING AND HYPERTENSION

| FUNCTIONAL MEASURE | EXPERIMENTAL LV PRESSURE LOADING | NORMOTENSIVE AGING |
|---|---|---|
| Contraction Duration | ↑ | ↑ |
| Contraction Velocity | ↓ | ↓ |
| Myosin Isozyme Composition | ↓$V_1$, ↑$V_3$ | ↓$V_1$, ↑$V_3$ |
| Sarcoplasmic Reticulum $Ca^{2+}$ Pumping Rate | ↓ | ↓ |
| Cytosolic $Ca^{2+}$ Transient Duration | ↑ (Ferret) | ↑ |
| Myofilament $Ca^{2+}$ Sensitivity | ↔ | ↔ |
| Action Potential Repolarization Time | ↑ | ↑ |
| β-Adrenergic Inotropic Response | ↓ | ↓ |
| Cardiac Glycoside Response | ↓ | ↓ |

Figure 10: The phenotypic *pattern* of experimental pressure loading in young animals bears a striking resemblance to that of adult aging and suggests that the molecular mechanisms may be similar in both cases.

endothelium may lead to the production of growth factors which can initiate protein synthesis. There is also evidence for an interaction of stretch and hormone receptor mediated intracellular signal transduction, e.g., cAMP or phosphatidylinositol turnover (134,142) or ion flux, e.g., $Na^+$ (138), possibly related to stretch induced activation of ionic channels (143).

The initial events in response to increased aortic pressure mimic the normal stress and growth responses and include the transient induction of heat shock proteins (144-146) and protooncogenes (139,147-149). There is also an induction of proteins that are usually present in ventricular myocardium during the fetal period, e.g., alpha skeletal actin (147), beta-tropomyosin (147) or of proteins that are usually present only in the atria of adult hearts, e.g., AnF (151,152). Subsequently, changes in gene coding for myofilament, channel or pump proteins occur. Initially at least, substantial spatio-temporal heterogeneity exists in the increase in mRNA encoding these various proteins (153,154). This could, in part, be due to heterogeneity of the specific factors stimulating the initial growth response. The phenotype of the stable, adapted hypertensive heart (Fig. 10) in some cases is due to differential expression of multigene families of contractile proteins, e.g., the shift of the myosin heavy chain isoform, and in other instances involves differential activation of single genes. In some cases the degree of activation of single genes or synthesis of gene products fails to keep pace with the increase in cell size and results in the decrease in the density of a given protein, e.g., the SR $Ca^{2+}$ pump (51,155). In other instances, up regulation of a gene, e.g., that coding for the $Ca^{2+}$ channel proteins, maintains a normal or enhanced density (156). (While the mRNA for $Ca^{2+}$ channels in cardiac pressure overload has yet to be studied, evidence has been presented that $Ca^{2+}$ channel synthesis is increased to maintain a normal (38) or enhanced (37,45,126) density of $Ca^{2+}$ channels.)

# REFERENCES

1.  Gerstenblith G, Frederiksen J, Yin FCP, *et al.* Echocardiographic assessment of a normal adult aging population. *Circulation* 56: 273, 1977.
2.  Gardin JM, Henry WL, Savage DD, *et al.* Echocardiographic measurements in normal subjects: evaluation of an adult population without clinically apparent heart disease. *J Clin Ultrasound* 7: 439, 1979.
3.  Sjogren AL. Left ventricular wall thickness determined by ultrasound in 100 subjects without heart disease. *Chest* 60: 341, 1971.
4.  Lakatta EG. Alterations in the cardiovascular system that occur in advanced age. *Fed Proc* 38: 163, 1979.
5.  Devereux RB, Savage DD, Drayer JIM, Laragh JH. Left ventricular hypertrophy and function in high, normal, and low-renin forms of essential hypertension. *Hypertension* 4: 524, 1982.
6.  Drayer JIM, Weber MA, DeYoung JL. Blood pressure as a determinant of cardiac left ventricular muscle mass. *Arch Intern Med* 143: 90, 1983.
7.  Shkhvatsabaya IK, Usubaliyev NN, Yurenev AP, *et al.* The interrelations of cardiac and vascular wall hypertrophy in arterial hypertension. *Cardiovasc Rev Rep* 2: 1145, 1981.
8.  Gerstenblith G, Fleg JL, Becker LC, *et al.* Maximum left ventricular filling rate in healthy individuals measured by gated blood pool scans. Effect of age. *Circulation* 68: III-101 1983.
9.  Savage DD, Drayer JIM, Henry WL, *et al.* Echocardiographic assessment of cardiac anatomy and function in hypertensive subjects. *Circulation* 59: 623, 1979.
10. Wolinsky H. Long-term effects of hypertension on the rat aortic wall and their relation to concurrent aging changes. Morphological and chemical studies. *Circ Res* 30: 301, 1972.
11. Gerstenblith G, Lakatta EG, Weisfeldt ML. Age changes in myocardial function and exercise response. *Prog Cardiovasc Res* 19: 1, 1976.
12. Bader H. Dependence of wall stress in the human thoracic aorta on age and pressure. *Circ Res* 30: 354, 1967.
13. Learoyd BM, Taylor MG. Alterations with age in the viscoelastic properties of human arterial walls. *Circ Res* 18: 278, 1966.
14. Yin FCCP, Spurgeon HA, Kallman CH. Age-associated alterations in viscoelastic properties of canine aortic strips. *Circ Res* 53: 464, 1963.
15. Landowne M. The relation between intra-arterial pressure and impact pulse wave velocity with regard to age and arteriosclerosis. *J Gerontol* 13: 153, 1958.
16. Avolio AP, Fa-Quan D, We-Qiang L, *et al.* Effects of aging on arterial distensibility in populations with high and low prevalence of hypertension: comparison between urban and rural communities in China. *Circulation* 71: 202, 1985.
17. Ting CT, Brin KP, Lin SJ, *et al.* Arterial hemodynamics in human hypertension. *J Clin Invest* 78: 1462, 1986.

18. O'Rourke MF. *Arterial Function in Health and Disease*. New York: Churchill Livingstone, p 1, 1982.

19. Gribbin B, Pickering TG, Sleight P, *et al*. Effect of age and high blood pressure on baroreflex sensitivity in man. *Circ Res* 29: 424 1971.

20. Bartel O, Buhler FR, Klowski W, *et al*. Decreased beta-adreno-receptor responsiveness as related to age, blood pressure, and plasma catecholamines in patients with essential hypertension. *Hypertension* 2: 130, 1980.

21. Kuramoto K, Matsushita S, Mifune J, *et al*. Electrocardiographic and hemodynamic evaluation of isoproterenol test in elderly ischemic heart disease, *Jpn Circ J* 42: 955, 1978.

22. Lakatta EG. Age-related alterations in the cardiovascular response to adrenergic mediated stress. *Fed Proc* 39: 3173, 1980.

23. London DM, Safar ME, Weiss YA, *et al*. Isoproterenol sensitivity and total body clearance of propranolol in hypertensive patients. *J Clin Pharmacol* 16: 174 1976.

24. Yin FCP, Spurgeon HA, Greene HL, *et al*. Age-associated decrease in heart rate response to isoproterenol in dogs. *Mech Aging Dev* 10: 17, 1979.

25. Yin FCP, Spurgeon HA, Raizes GS, *et al*. Age associated decrease in chronotropic response to isoproterenol. *Circulation* 54: II-167, 1976.

26, Ibsen H, Julius S. Pharmacologic tools for assessment of adrenergic nerve activity in human hypertension. *Fed Proc* 43: 67, 1984.

27. Lakatta EG, Yin FCP. Myocardial aging: Functional alterations and related cellular mechanisms. *Am J Physiol* 242: H927, 1982.

28. Ibsen H, Julius S. Pharmacologic tools of assessment of adrenergic nerve activity in human hypertension. *Fed Proc* 43: 67, 1984.

29. Dzau VJ, Safar ME. Large conduit arteries in hypertension: role of the vascular renin-angiotensin system. *Circulation* 77: 947, 1988.

30. Avolio AP, Chen S-G, Wang R-P, *et al*. Effects of aging on changing arterial compliance and left ventricular load in a northern Chinese urban community. *Circulation* 68: 50, 1983.

31. Lakatta EG. Chaotic behavior of myocardial cells: possible implications regarding the pathophysiology of heart failure. *Perspect Biol Med* 32: 421, 1989.

32. Capasso JM, Malhotra A, Scheuer J, Sonnenblick EH. Myocardial biochemical, contractile and electrical performance following imposition of hypertension in young and old rats. *Circ Res* 58: 445, 1986.

33. Gulch RW, Baumann R, Jacob R. Analysis of myocardial action potential in left ventricular hypertrophy of Goldblatt rats. *Basic Res Cardiol* 74: 69, 1979.

34. Aronson RS. Characteristics of action potential of hypertrophied myocardium from rats with renal hypertension. *Circ Res* 47: 443, 1980.

35. Capasso JM, Aronson RS, Sonnenblick EH. Reversible alterations in excitation-contraction coupling during myocardial hypertrophy in papillary muscle. *Circ Res* 51: 189, 1982.

36. Keung ECH, Aronson RS. Non-uniform electrophysiological properties and electrotonic interaction in hypertrophied rat myocardium. *Circ Res* 49: 150 1981.

37. Keung EC. Calcium current is increased in isolated adult myocytes from hypertrophied rat myocardium. *Circ Res* 64: 753, 1989.

38. Mayoux E, Scamps F, Oliviero P, Vasson G, Charlemagne D. Calcium channels in normal and hypertrophied rat heart. *J Mol Cell Cardiol* 21: S-18, 1989.

39. Mayoux E, Callens F, Swynghedauw B , Charlemagne D. Adaptational process of the cardiac Ca$^{2+}$ channels to pressure overload: biochemical and physiological properties of the dihydropyridine receptors in normal and hypertrophied rat hearts. *J Cardiovasc Pharmac* 12: 390, 1988.

40. Lecarpentier Y, Bugaisky LB, Chemla D, Mercadier JJ, Schwartz K, Whalen RG, Martin JL. Coordinated changes in contractility, energetics, and isomyosins after aortic stenosis. *Am J Physiol* 252: H282, 1987.

41. Bassett AL, Gelband H. Chronic partial occlusion of the pulmonary artery in cars. *Circ Res* 32: 15, 1973.

42. Teneick RE, Houser SR, Bassett AL. Cardiac hypertrophy and altered cellular electrical activity of the myocardium: possible electrophysiological basis for myocardial contractility changes. *In: Physiology and Pathophysiology of the Heart*, 2nd edition, N. Sperelakis (ed). Norwell: Kluwer Academic Publishers, p 573, 1989.

43. Tritthart H, Luedcke H, Bayer R, Stierle H, and Kaufmann R. Right ventricular hypertrophy in the Cat--an electrophysiological and anatomical study. *J Mol Cell Cardiol* 7: 163,, 1975.

44. Kaufmann RL, Homburger H, Wirth H. Disorder in excitation-contraction coupling of cardiac muscle from cats with experimentally produced right ventricular hypertrophy. *Circ Res* 58: 346 1971.

45. Kleiman RB, Houser SR. Calcium currents in normal and hypertrophied isolated feline ventricular myocytes. *Am J Physiol* 255: H1424, 1988.

46. Kleiman RB, Houser SR. Outward currents in normal and hypertrophied feline ventricular myocytes. *Am J Physiol* 256: H1450, 1989.

47. Gwathmey JK, Morgan JP. Altered calcium handling in experimental pressure-overload hypertrophy in the ferret. *Circ Res* 57: 837, 1985.

48. Dinda BB, Carson NL, Houser SR. Calcium transient and contractile properties of feline ventricular myocytes in heart failure. *FASEB J* 3:A984, 1989.

49. Morgan JP, MacKinnon R, Feldman M, Grossman W, Gwathmey J. The effects of cardiac hypertrophy on intracellular Ca$^{2+}$ handling. *In: Diastolic Relaxation of the Heart*, W. Grossman and B. Lorell (eds). Boston: Martinus Nijhoff, p 97, 1988.

50. Boyett MR, Capogrossi MC, duBell WH, Lakatta EG, Spurgeon HA. Cytosolic Ca$^{2+}$ modulation of the action potential in right ventricular myocytes. *J Physiol (Lond)* 415: 109P, 1989.

51. de la Bastie D, Levitsky D, Rappaport L, *et al.* Function of the sarcoplasmic reticulum and expression of its Ca$^{2+}$ ATPase gene in pressure overload-induced cardiac hypertrophy in the rat. *Circ Res* 66: 554, 1990.

52. Lamers JMJ, Stinis JT. Defective calcium pump in the sarcoplasmic reticulum of the hypertrophied rabbit heart. *Life Sci* 24: 2313 1979.

53. Alpert NR, Blanchard EM, Mulieri LA. The quantity and rate of Ca$^{2+}$ uptake in normal and hypertrophied hearts. *In: Pathophysiology of Heart Disease,* N. Dalla, P. Singel, R.E. Bemish (eds), New York: Raven Press, p 99, 1987.

54. Nagai R, Zarain-Herzberg A, Brandl CJ, *et al.* Regulation of myocardial Ca$^{2+}$-ATPase and phospholamban mRNA expression in response to pressure overload and thyroid hormone. *Proc Natl Acad Sci USA* 86: 2966, 1989.

55. Charlemagne D, Maixen J-M, Preteseille M, Lelievre LG. Ouabain binding sites and (Na+,K+)-ATPase activity in rat cardiac hypertrophy. *J Biol Chem* 261: 185, 1986.

56. Orlowski J, J. B. Lingrel. Differential expression of the Na,K-ATPase $\alpha_1$ and $\alpha_2$ subunit genes in murine myogenic cell line. *J Biol Chem* 263: 17817, 1988.

57. Houser SR, Freeman AR, Jaeger JM, Breisch EA, Coulson RL, Carey R, Spann JF. Resting potential changes associated with Na-K pump in failing heart muscle. *Am J Physiol* 240: H168, 1981.

58. Hanf R, Drubaix I, Marotte F, Lelievre LG. Rat cardiac hypertrophy. Altered sodium-calcium exchange activity in sarcolemma vesicles. *FEB* 236: 145, 1988.

59. Fabiato A, Fabiato F. Excitation-contraction coupling of isolated cardiac fibers with disrupted or closed sarcolemma; calcium dependent cyclic and tonic contractions. *Circ Res* 32: 293, 1972.

60. Kass RS, Tsien RW, Weingart R. Ionic basis of transient inward current induced by strophanthidin in cardiac Purkinje fibres. *J Physiol (Lond)* 281: 209, 1978.

61. Lakatta EG, Dappe DL. Diastolic scattered light fluctuation, resting force and twitch force in mammalian cardiac muscle. *J Physiol (Lond)* 315: 369 1981.

62. Stern MD, Kort AA, Bhatnagar GM, Lakatta EG. Scattered-light intensity fluctuations in diastolic rat cardiac muscle caused by spontaneous $Ca^{2+}$-dependent cellular mechanical oscillations. *J Gen Physiol* 82: 119, 1983.

63. Kort AA, Lakatta EG. Calcium-dependent mechanical oscillations occur spontaneously in unstimulated mammalian cardiac tissues. *Circ Res* 54: 119, 1983.

64. Lakatta EG, Capogrossi MC, Kort AA, Stern MD. Spontaneous myocardial Ca oscillations: an overview with emphasis on ryanodine and caffeine. *Fed Proc* 44: 2977, 1985.

65. Lakatta EG, Capogrossi MC, Spurgeon HA, Stern MD. Characteristics and functional implications of spontaneous sarcoplasmic reticulum generated cytosolic calcium oscillations in myocardial tissue. *In: Cell Calcium Metabolism: Physiology, Biochemistry, Pharmacology and Clinical Implications*, G. Fiskum (ed). New York: Plenum Publishing Co., p 529, 1989.

66. Capogrossi MC, Kort AA, Spurgeon HA, Lakatta EG. Single adult rabbit and rat cardiac myocytes retain the $Ca^{2+}$- and species-dependent systolic and diastolic properties of intact muscle. *J Gen Physiol* 88: 589, 1986.

67. Kort AA, Capogrossi MC, Lakatta EG. Frequency, amplitude, and propagation velocity of spontaneous $Ca^{2+}$-dependent contractile waves in intact adult rat cardiac muscle and isolated myocytes. *Circ Res* 57: 844, 1985.

68. Wier WG, Kort AA, Stern MD, *et al.* Cellular calcium fluctuations in mammalian heart: direct evidence from noise analysis of aequorin signals in Purkinje fibers. *Proc Natl Acad Sci USA* 80: 7367, 1983.

69. Orchard CH, Eisner DA, Allen DG. Oscillations of intracellular $Ca^{2+}$ in mammalian cardiac muscle. *Nature* 30: 735, 1983.

70. Capogrossi MC, Lakatta EG. Frequency modulation and synchronization of spontaneous oscillations in cardiac cells. *Am J Physiol* 248: H412, 1985.

71. Capogrossi MC, Suarez-Isla BA, Lakatta EG. The interaction of electrical stimulated twitches and spontaneous contractile waves in single cardiac myocytes. *J Gen Physiol* 88: 615, 1986.

72. Capogrossi MC, Stern MD, Spurgeon HA, Lakatta EG. Spontaneous Ca$^{2+}$ release from the sarcoplasmic reticulum limits Ca$^{2+}$-dependent twitch potential in individual cardiac myocytes: a mechanism for maximum inotropy in the myocardium. *J Gen Physiol* 91: 133, 1988.

73. Kort AA, Lakatta EG. The relationship of spontaneous sarcoplasmic reticulum calcium release in twitch tension in rat and rabbit cardiac muscle. *Circ Res* 63: 969, 1988.

74. Kass RS, Tsien RW. Fluctuations in membrane current driven by intracellular calcium in cardiac Purkinje fibers. *Biophys J* 38: 259, 1982.

75. Lederer WJ, Tsien RW. Transient inward current underlying arrhythmogenic effects of cardiotonic steroids in Purkinje fibres. *J Physiol (Lond)* 263: 73, 1976.

76. Ferrier GR, Saunders JH, Mendez CA. Cellular mechanism for the generation of ventricular arrhythmias by acetylstrophanthydine. *Circ Res* 32: 600, 1973.

77. Capogrossi MC, S. Houser, A. Bahinski, Lakatta EG. Synchronous occurrence of spontaneous localized calcium release from the sarcoplasmic reticulum generates action potentials in rat cardiac ventricular myocytes at normal resting membrane potential. *Circ Res* 61: 498, 1987.

78. Aronson RS. Afterpotentials and triggered activity in hypertrophied myocardium from rats with renal hypertension. *Circ Res* 48:720, 1981.

79. Heller LJ. Augmented aftercontractions in papillary muscles from rats with cardiac hypertrophy, *Am J Physiol* 6: H649, 1979.

80. J. A. Shechter, Friehling TD, Uboh C, Kelliher GJ, O'Connor KM, Kowey PR. The effect of left ventricular hypertrophy on inducible ventricular arrhythmias. *Circulation* 70: II-224, 1984.

81. January CT, Riddle JM, Salata JJ. A model for early afterdepolarizations: induction with the Ca$^{2+}$ channel agonist Bay K 8644. *Circ Res* 62: 563, 1988.

82. Spurgeon HA, duBell W, Boyett M, Talo A, Capogrossi MC, Lakatta EG. Cytosolic Ca$^{2+}$ modulation of membrane potential during a heart beat: perspectives from the single cardiac cell. *J Mol Cell Cardiol* 21: S-19, 1989.

83. Anversa P, Lond V, Giacomelli F, Weiner J. Absolute morphometric study of myocardial hypertrophy in experimental hypertension. II. Ultrastructure of myocytes and interstitium. *Lab Invest* 38: 597, 1978.

84. Jacob R, Kissling G, Ebrecht G, Holubarsch C, Medugorac I, Rupp H. Adaptive and pathological alterations in experimental cardiac hypertrophy. *In: Advances in Myocardiology,* Volume IV, E. Chazov, V. Saks, and G. Rona (eds). New York: Plenum Publishing Co., p 55, 1983.

85. Alpert NR, Mulieri LA. Increased myothermal economy of isometrical force generation in compensated cardiac hypertrophy induced by pulmonary artery constriction in the rabbit. Characterization of heat liberation in normal and hypertrophied right ventricular papillary muscles. *Circ Res* 50: 491, 1982.

86. Lecarpentier Y, Waldenstrom A, Clerque M, Chemia D, Oliviero P, Martin JL, Swynghedauw B. Major alterations in relaxation during cardiac hypertrophy induced by aortic stenosis in guinea pig. *Circ Res* 61: 107, 1987.

87. Ebrecht G, Rupp H, Gacob R. Alterations of mechanical parameters in chemically skinned preparations of rat myocardium as a function of isoenzyme pattern of myosin. *Basic Res Cardiol* 77: 220, 1982.

88. Ventura-Clapier R, Mekhfi H, Oliviero P, Swynghedauw B. Pressure overload changes cardiac skinned-fiber mechanics in rats, not in guinea pigs. *Am J Physiol* 254: H517, 1988.

89. Henry PD, Ahumada GG, Friedman NF, Sobel BC. Simultaneously measured isometric tension and ATP hydrolysis in glycerinated fibres from normal and hypertrophied rabbit heart. *Circ Res* 31: 740, 1972.

90. Maughan D, Low E, Litten R, Brayden J, Alpert N. Calcium activated muscle from hypertrophied rabbit hearts. Mechanical and correlated biochemical changes. *Circ Res* 44: 279, 1979.

91. Scheuer J, Bhan AK. Cardiac contractile proteins. Adenosine triphosphatase activity and physiological function. *Circ Res* 45: 1, 1979.

92. Shiverick KT, Hamrell BB, Alpert NR. Structural and functional properties of myosin associated with the compensatory cardiac hypertrophy in the rabbit. *J Mol Cell Cardiol* 8: 837, 1976.

93. Wisenbaugh T, Allen P, Copper G IV, *et al.* Hypertrophy without contractile dysfunction after reversal of pressure overload in the cat, *Am J Physiol* 247: H146, 1984.

94. Afflitto JJ, Inchiosa MA Jr. Decrease in rat cardiac myosin ATPase with aortic constriction: prevention by thyroxine treatment. *Life Sci* 25: 353, 1979.

95. Hoh JFY, Rossmanith GH. Ventricular isomyosins and the tonic regulation of cardiac contractility. *In: Pathobiology of Cardiovascular Injury*, H.L. Stone and W.B. Weglicki (eds). Boston: Martinus Nijhoff, p 476, 1985.

96. Mercadier J-J, Lompre A-M, Wisnewsky C, Samuel J-L, Bercovici J, Swynghedauw B, Schwartz K. Myosin isoenzymic changes in several models of rat cardiac hypertrophy. *Circ Res* 49: 525, 1981.

97. Scheuer J, Malhotra A, Hirsch G, Capasso J, Schaible TF. Physiologic cardiac hypertrophy corrects contractile protein abnormalities associated with pathologic hypertrophy in rats. *J Clin Invest* 70: 1300, 1982.

98. Rupp H. The adaptive changes in the isoenzyme pattern of myosin from hypertrophied rat myocardium as a result of pressure overload and physical training. *Basic Res Cardiol* 76: 79, 1981.

99. Lompre AM, Schwartz K, Albis A, Lacombe G, Thiem NV, Swynghedauw B. Myosin isozymes redistribution in chronic heart overloading. *Nature* 282: 105, 1979.

100. Mercadier J-J, Bouveret P, Gorza L, Schiaffino S, Clark WA, Zak R, Swynghedauw B, Schwartz K. Myosin isoenzymes in normal and hypertrophied human ventricular myocardium. *Circ Res* 53: 52, 1983.

101. Lompre AM, Mercadier JJ, Wisnewsky C, *et al.* Species- and age-dependent changes in relative amounts of cardiac myosin isozymes in mammals. *Dev Biol* 84: 286, 1981.

102. Gorza L, Mercadier JJ, Schwartz K, Thornell LE, Sartore S, Schiaffino S. Myosin types in the human heart. An immunofluorescence study of normal and hypertrophied atrial and ventricular myocardium. *Circ Res* 54: 694, 1984.

103. Julian FJ, Mogan DL, Moss RL, Gonzalez M, Dwivedi P. Myocyte growth without physiological impairment in gradually induced rat cardiac hypertrophy. *Circ Res* 49: 1300, 1981.

104. Lecarpentier YC, Chuck LHS, Housmans PR, DeClerck NM, Brutsaert DL. Nature of load dependence of relaxation in cardiac muscle. *Am J Physiol* 237: H460, 1979.

105. O'Rourke MF. *Arterial Function in Health and Disease.* New York: Churchill Livingstone, 1982.
106. Chevalier B, Mansier P, Amrani FC-E, Swynghedauw B. β-adrenergic system is modified in compensatory pressure cardiac overload in rats; physiological and biochemical evidence. *J Cardiovasc Pharmac* 13: 412, 1989.
107. Mansier P, Chevalier B, Swynghedauw B. Characterization of the beta adrenergic system in adult rat hypertrophied hearts. *J Mol Cell Cardiol* 21: S-17, 1989.
108. Ayobe HM, Tarazi CC. Reversal of changes in myocardial β-receptor and inotropic responsiveness with regression of cardiac hypertrophy in renal hypertensive rats (RHR). *Circ Res* 54: 125, 1984.
109. Gende OA, Mattiazzi A, Million MC, *et al.* Renal hypertension impairs inotropic isoproterenol effect without β-receptor changes. *Am J Physiol* 249: 814, 1985.
110. Woodcock WA, Funder JW, Johnston CI. Decreased cardiac β-adrenergic receptors in deoxycorticosterone-salt and renal hypertensive rats. *Circ Res* 45: 560, 1979.
111. Schwartz K, Lompre AM, De la Bastie D, Mercadier JJ. Mechanogenic transduction in the hypertrophied heart. *J Mol Cell Cardiol* (Suppl III) 21: S-24, 1989.
112. Weber KT. Angiotensin II and myocardial remodeling, *J Mol Cell Cardiol* (Suppl III) 21: S-28, 1989.
113. Fraticelli A, Josephson R, Danziger R, Lakatta E, Spurgeon H. Morphological and contractile characteristics of rat cardiac myocytes from maturation to senescence. *Am J Physiol* 257: H259, 1989.
114. Capasso JM, A. Malhorta, Remily RM, Scheuer J, Sonnenblick EH. Effects of age on mechanical and electrical performance of rat myocardium. *Am J Physiol* 245: H72, 1983.
115. Wei JY, Spurgeon HA, Lakatta EG. Excitation-contraction in rat myocardium: alterations with adult aging. *Am J Physiol* 246: H784, 1984.
116. Orchard CH, Lakatta EG. Intracellular calcium transients and developed tensions in rat heart muscle. A mechanism for the negative interval-strength relationship. *J Gen Physiol* 86: 637, 1985.
117. Lakatta EG, Yin FCP. Myocardial aging: Functional alterations and related cellular mechanisms, *Am J Physiol* 242: H927, 1982.
118. Alpert NR, Gale HH, Taylor N. The effect of age on contractile protein ATPase activity and the velocity of shortening. *In: Factors influencing myocardial contractility*, R.D. Tanz, F. Kavaler, and J. Roberts (eds). New York: Academic, p 127, 1967.
119. Effron MB, Bhatnagar GM, Spurgeon HA, Ruano-Arroyo G, Lakatta EG. Changes in myosin isoenzymes, ATPase activity, and contraction duration in rat cardiac muscle with aging can be modulated by thyroxine. *Circ Res* 60: 238, 1987.
120. Bhatnagar GM, Walford GD, Beard ES, Humphreys SH, Lakatta EG. ATPase activity and force production in myofibrils and twitch characteristics in intact muscle from neonatal, adult, and senescent rat myocardium. *J Mol Cell Cardiol* 16: 203, 1984.
121. Froehlich JP, Lakatta EG, E. Beard, Spurgeon HA, Weisfeldt ML, Gerstenblith G. Studies of sarcoplasmic reticulum function and contraction duration in young and aged rat myocardium. *J Mol Cell Cardiol* 10: 427, 1978.
122. Guarnieri T, Filburn CR, Beard ES, Lakatta EG. Enhanced contractile response and protein kinase activation to threshold levels

of β-adrenergic stimulation in hyperthyroid rat heart. *J Clin Invest* 65: 861, 1980.

123. Lakatta EG, Gerstenblith G, Angell CS, Shock NW, Weisfeldt ML. Diminished inotropic response of aged myocardium to catecholamines. *Circ Res* 36: 262, 1975.

124. Lakatta EG, Gerstenblith G, Angell CS, Shock NW, Weisfeldt ML. Prolonged contraction duration in aged myocardium. *J Clin Invest* 55: 61, 1975.

125. Spurgeon HA, Steinbach MF, Lakatta EG. Chronic exercise prevents characteristic age-related changes in rat cardiac contraction. *Am J Physiol* 244: H513 1983.

126. Walker KE, Lakatta EG, Houser SR. Calcium currents in senescent rat ventricular myocytes. *Circulation* 80: II-142, 1989.

127. Narayanan N. Differential alterations in ATP-supported calcium transport activities of sarcoplasmic reticulum and sarcolemma of aging myocardium. *Biochim Biophys Acta* 678: 442 1981.

128. Guarnieri T, Filburn CR, Zitnik G, Roth GS, Lakatta EG. Contractile and biochemical correlates of β-adrenergic stimulation of the aged heart. *Am J Physiol* 239: H501, 1980.

129. Weisfeldt ML, Lakatta EG, Gerstenblith G. Aging and cardiac disease. *In: Heart Disease: A Textbook of Cardiovascular Medicine,* 3rd ed., E. Braunwald (ed). Philadelphia: W.B. Saunders, p 1650, 1988.

130. Sakai M, Danziger RS, Spurgeon HA, Lakatta EG. Decreased contractile response to norepinephrine with aging. *Circulation* 76: IV-153, 1987.

131. Sakai M, Danziger RS, Staddon JM, Lakatta EG, Hansford RG. Decrease with senescence in the norepinephrine-induced phosphorylation of myofilament proteins in isolated rat cardiac myocytes. *J Mol Cell Cardiol* 21:1327, 1989.

132. Cooper G IV, Mercer WE, Hoober JK, *et al.* Load regulation of the properties of adult feline cardiocytes. Role of substrate adhesion. *Circ Res* 58: 692, 1986.

133. Bauters C, Moalic J-M, Bercovici J, *et al.* Augmentation de l'expression des oncogenenes *c-myc* et *c-fos* en fonction de l'activite mecanique du coeur isole de rat adulte. *CR Acad Sci Paris* (Serie III) 306: 597, 1988.

134. Watson PA, Haneda T, Morgan HE. Effect of higher aortic pressure on ribosome formation and cAMP content in rat heart. *Am J Physiol* 256: C1257, 1989.

135. Simpson P. Stimulation of hypertrophy of cultured neonatal rat heart cells through an $\alpha_1$- and $\beta_1$-adrenergic receptor interaction. Evidence for independent regulation of growth beating. *Circ Res* 56: 884, 1985.

136. Mann DL, Kent RL, Cooper G IV. Load regulation of the properties of adult feline cardiocytes: growth induction by cellular deformation. *Circ Res* 64: 1079, 1989.

137. Yazaki Y, Komuro I. Molecular analysis of cardiac hypertrophy due to overload. *J Mol Cell Cardiol* (suppl III) 21: S-29, 1989.

138. Kent RL, Hoober JK, Cooper G IV. Load responsiveness of protein synthesis in adult mammalian myocardium: role of cardiac deformation linked to sodium influx. *Circ Res* 64: 74, 1989.

139. Bauters C, Moalic JM, Bercovici J, *et al.* Coronary flow as a determinant of *c-myc* and *c-fos* proto-oncogene expression in an isolated adult rat heart. *J Mol Cell Cardiol* 20: 97, 1988.

140. Lee HR, Henderson SA, Reynolds R, Dunnmon P, Yuan D, and Chien KR. $\alpha_1$-adrenergic stimulation of cardiac gene transcription in neonatal rat myocardial cells. *J Biol Chem* 263: 7352, 1988.

141. Watkins SC, Samuel JL, Marotte F, Bertier-Savalle B, Rappaport L. Microtubules and desmin filaments during onset of heart hypertrophy in rat: a double immunoelectron microscope study. *Circ Res* 60: 327, 1987.

142. Von Harsdorf R, Lang RE, Fullerton M, Woodcock EA. Myocardial stretch stimulates phosphatidylinositol turnover. *Circ Res* 65: 494, 1989.

143. Craelius W, Chen V, El-Sherif N. Stretch activated ion channels in ventricular myocytes. *Biosci Rep* 8: 407, 1988.

144. Delcayre C, Samuel J-L, Marotte F, Best-Belpomme M, Mercadier JJ, Rapport L. Synthesis of stress proteins in rat cardiac myocytes 2-4 days after imposition of hemodynamic overload. *J Clin Invest* 82: 460, 1988.

145. Simpson PC. Molecular mechanisms in myocardial hypertrophy. *Heart Failure* 5: 113, 1989.

146. Hammon GL, Lai YK, Markert CL. Diverse forms of stress lead to new patterns of gene expression through a common and essential metabolic pathway. *Proc Natl Acad Sci USA* 79: 3485, 1982.

147. Izumo S, Nadal-Ginard B, Mahdavi V. Protooncogene induction and reprogramming of cardiac gene expression produced by pressure overload. *Proc Natl Acad Sci USA* 85: 339, 1988.

148. Simpson PC. Proto-oncogenes and cardiac hypertrophy. *Annu Rev Physiol* 51: 189, 1988.

149. Mulvagh SL, Michael LH, Perryman MB, Roberts R, Schneider MD. A hemodynamic load *in vivo* induces cardiac expression of the cellular oncogene, *c-myc*. *Biochem Biophys Res Commun* 147: 627, 1987.

150. Schwartz K, De la Bastie D, Bouveret P, Oliviero P, Alonso S, Buckingham M. $\alpha_1$ skeletal muscle actin mRNA's accumulate in hypertrophied adult rat hearts. *Circ Res* 59: 551, 1986.

151. Seidman CE. Expression of atrial natriuretic factor in the normal and hypertrophied heart. *Heart Failure* 5: 130, 1989.

152. Day ML, D. Schwartz, Wiegand RC, Stockman PT, Brunnert SR, Tolunay HE, Currie MG, Standaert DG, Needleman P. Ventricular atriopeptin. unmasking of messenger RNA and peptide synthesis by hypertrophy or dexamethasone. *Hypertension* 9: 485, 1987.

153. Rappaport L, Contard F, Marotte F, Mebazza A, Delcayre C, Samuel JL. Regional distribution of growth signals, contractile and extracellular matrix proteins within myocardium following the imposition of a sudden pressure overload. *J Mol Cell Cardiol* (Suppl III) 21: S-22, 1989.

154. Schiaffino S, Samuel JL, D. Sassoon, *et al.* Non-synchronous accumulation of $\alpha_1$ skeletal actin and $\beta$-myosin heavy chain mRNAs during early stages of pressure overload-induced cardiac hypertrophy demonstrated by in situ hybridization. *Circ Res* 64: 937, 1989.

155. Dillmann WH, Rohrer D, Maciel L. Age induced decrease in the mRNA coding for $Ca^{++}$ ATPase of the sarcoplasmic reticulum of the rat heart. *Clin Res* 37: 516A, 1989.

156. Swynghedauw B. Remodeling of the heart in response to chronic mechanical overload. *Eur Heart J* 10: 935, 1989.

# LEFT VENTRICULAR HYPERTROPHY: DISSOCIATION OF

# STRUCTURAL AND FUNCTIONAL EFFECTS BY THERAPY

Edward D. Frohlich

Vice President for Academic Affairs
Alton Ochsner Medical Foundation
New Orleans, LA 70120

The heart is one of the major target organs that becomes secondarily involved with the unrelenting and progressive disease of essential hypertension (1,2). As a result of this increasing afterload that is imposed upon the left ventricle, the ventricular chamber adapts structurally as well as functionally. The functional changes involve the enhanced requirements of the ventricle that permits the chamber to overcome the pressure, and possibly also, the volume overload that is associated with the disease (3-6). The structural changes involve an increase in muscle mass that is achieved through the process of left ventricular hypertrophy (LVH) in a manner that may be similar to the arteriolar changes demonstrated by an increased wall-to-lumen ratio that is produced by the increased thickening of the wall (5-7).

Unless antihypertensive therapy is interdicted in the disease process, left ventricular failure will ensue as the major cardiac hemodynamic consequence. However, left ventricular hypertrophy is also associated with a risk that is independent of the systolic (and diastolic) pressure overload and the related hemodynamic risks (8-10). Although antihypertensive therapy will reduce risk from the hemodynamic alterations, only recently have epidemiological findings suggested that the independent risk of the LVH may be reduced with pharmacological therapy (11), but there are no data presently available to indicate just which agents will reduce that risk (12). All long-term antihypertensive treatment programs may be expected to reduce LVH, only certain classes of agents will do so within a few weeks and independent of the hemodynamic factors (4-6,13,14). Some of these agents may even impair cardiac and other hemodynamic functions if the arterial pressure is increased abruptly following therapeutic reduction of cardiac mass (15,16), whereas other antihypertensive agents have been shown to preserve normal cardiac function or may even improve cardiac performance (17,18).

This review concerns the clinical and experimental evidence supporting the classical concept that functional overload is responsible for the structural adaptation of the left ventricle in hypertension. In addition,

*Cellular and Molecular Mechanisms in Hypertension*
Edited by R.H. Cox, Plenum Press, New York, 1991

newer evidence will be presented to support the hypothesis that non-hemodynamic factors are also responsible for the development of LVH in hypertension (2-6,19,20). Included in this latter information will be recent data which indicate that with specific antihypertensive pharmacological agents, reduction of left ventricular mass may be dissociated from their hemodynamic effects (14,17-20). For the most part, discussion concerning clinical and experimental forms of hypertension will be related to patients with essential hypertension and with the genetically-derived spontaneously hypertensive rat.

## ESSENTIAL HYPERTENSION AND THE SPONTANEOUSLY HYPERTENSIVE RAT

Essential hypertension is the most common form of hypertension in man, occurring in approximately 95 percent of all patients with hypertensive disease. The disease is generally considered to be produced by a dysregulation of the myriad of factors that participate in the normal regulation of arterial pressure that is most likely produced by one or more genetically transmitted factors (5,21). In an effort to reproduce this as yet poorly undefined multifactorial disease, the spontaneously hypertensive rat (SHR) was developed by genetic brother-to-sister inbreeding over 25 years ago, and it remains as the best experimental laboratory model for the clinical disease (22,23).

Both the essential hypertensive patient and the experimental SHR rat demonstrate a persistent and progressive increase of arterial pressure of the natural history of the disease that reflects an increasing total peripheral resistance. Moreover, in both forms of hypertension left ventricular mass increases with the progression of the vascular disease and the pressure overload imposed upon the left ventricle (1,2). This structural adaptation to this increasing afterload is one of a concentric LVH; but unlike other clinical and experimental models of hypertensive disease (e.g., renal arterial disease or its experimental model, two-kidney, one-clip Goldblatt hypertension) the correlation between the mass of the left ventricle and the height of arterial pressure is very poor (4,5,13,14). In fact, in the SHR of a very wide age range the correlation between these two indices are statistically insignificant (24). This finding, therefore, suggested that other non-hemodynamic factors may also participate in the development and maintenance of LVH in these two forms of hypertension (2-5,19). Among those non-hemodynamic factors that have been identified are: the age and gender of the hypertensive patient or rat; the race of the patient; coexisting diseases in man (e.g., exogenous obesity, atherosclerotic vascular disease, diabetes mellitus); antihypertensive and other therapeutic agents; participation of circulating humoral substances and other growth factors; and other recent considerations including the integrated pressure elevation over a full 24-hour period (4-6).

In order to approach these recent concerns systematically and to understand these confounding factors related to the development and reversal of LVH, this discussion will be concerned primarily with controlled experimental studies involving the SHR. However, to support their clinical relevance we shall refer to supporting findings in the patient with essential hypertension.

# PHARMACOLOGICAL REVERSAL OF LVH IN THE SHR

## Clinical Problem

The following experimental studies were designed to answer the important concern that could be raised from clinical circumstances involving long-term antihypertensive therapy. Thus, as already stated, LVH is brought about as an adoptive phenomenon by the progression of untreated hypertension and, with treatment, this increased left ventricular mass could be reduced significantly. Since LVH provides the structural support necessary to maintain cardiac function in the face of the increasing afterload and to prevent an earlier development of cardiac failure, we were interested to learn what would happen when antihypertensive therapy that reduced cardiac mass was discontinued abruptly and arterial pressure was increased abruptly.

## Experimental Design

As already indicated, any form of antihypertensive pharmacological therapy will reduce cardiac mass in the hypertensive patient or in the SHR if maintained long enough. However, in order to attempt to dissociate the hemodynamic consequences of reduced arterial pressure from the possible drug-related structural effects of these pharmacological agents we developed a protocol for all of our studies that mandates a short-term course of antihypertensive therapy in the SHR. Because our earlier studies demonstrated that reduction of left ventricular mass in older or in female SHR rats was less predictable, the following reproducible "bioassay" protocol was developed. Only 16 to 20 week old, male, SHR rats were selected for study; and their responses were compared with those in age- and sex-matched controls, the normotensive Wistar-Kyoto (WKY) rat that does not have LVH (provided that these rats are checked for possible genetically developed arteriovenous shunting) (25,26).

## Protocol

In this fashion, all rats (SHR and WKY) are first checked by a rat tail plethysmographic technique previously reported from our laboratory to substantiate that they, in fact, did have an elevated or normal blood pressure, respectively (27). Once this had been verified, the rats were allocated into two treatment groups, one receiving the active antihypertensive agent, the dose of which had previously been shown to reduce arterial pressure in the SHR. The other group received the pharmacologically and physiologically inert vehicle for the active agent. The two groups were then treated only for a three-week period. They were given their respective forms of treatment by gastric gavage once daily.

## Hemodynamics

At the conclusion of the three week treatment period, unanesthetized arterial pressure was measured in the conscious state after arterial and venous polyethylene catheters were introduced under light-ether anesthesia for pressure measurement and infusion of blood or other fluids as indicated (28,29). A tracheostomy was performed for controlled ventilation after a midsternal thoracotomy was performed for the necessary hemodynamic studies. In this fashion we determined the pumping ability of the left ventricle (30,31) and the *in vivo* and *in vitro* assessment of aortic distensibility which were obtained at the pharmacologically-induced hypotensive levels as well as when pressure is abruptly elevated to pretreatment levels using a snare that had been placed previously around the ascending aorta (15,16). Another catheter was placed into the left ventricle for end-diastolic pressure measurement; and a 2.5 mm diameter electromagnetic flow probe is placed around the ascending aorta for determining cardiac output (30,31). These measurements were made before and during the rapid (one minute) infusion of homologously-derived whole blood from donor rats of the same strain. The same procedure is followed in ten control rats "treated" only with the inert vehicle.

## Left Ventricular Pumping Ability

Function curves of the left ventricular pumping ability were constructed by plotting the left ventricular end-diastolic pressures (LVEDP) against stroke index or cardiac index during that one minute of whole blood infusion (15). As indicated, these curves were obtained at the pharmacologically reduced arterial pressure as well as when pressure is increased to pre-treatment levels by occluding the ascending aortic snare.

## Aortic Distensibility

*In vivo* aortic distensibility curves were constructed from the changes in stroke index plotted against changes in the aortic distensibility index (stroke index divided by arterial pulse pressure) (15). In addition, *in vitro* curves of aortic distensibility were determined by ligating a measured segment of thoracic aorta and then measuring changes in developed intravascular pressure as known volumes of whole blood were added to that vessel segment (17).

## Initial Findings

Following this standardized protocol we were able to determine that despite the fact that all antihypertensive agents that were administered to the SHR and WKY produced predictable hemodynamic actions, not all agents diminished cardiac or aortic mass similarly (15,16,29). In fact, in our first study two beta-adrenergic receptor blocking agents, propranolol or timolol, were given to SHR and WKY "breeder" parents in their drinking water (32). Thus, the conceived rat pups were subjected to the beta-blocker from conception, through *in utero* development, following birth through the maternal milk while suckling, and then in their own drinking tap water until 12 weeks of age when they were studied hemodynamically. At that time all pharmacologically-treated rats demonstrated a reduced heart rate, cardiac output, and LVEDP; however, they all developed hypertension with the same degree of arterial pressure elevation as those SHR rats given only tap water. The purpose of the study, to determine whether chronically reduced cardiac output from conception would prevent eventual development of hypertension, was not confirmed. The effects of these agents on hemodynamics and cardiac mass provided our first evidence of a dissociation between the functional and structural effects of pharmacological agents. Both beta-adrenergic receptor blocking agents reduced arterial pressure, but only in the male SHR rats; but neither agent produced a reduction in cardiac mass in these male rats. In contrast, however, both agents reduced cardiac mass of the female rats, but without reducing their arterial pressure (32).

## Initial "Bioassay" Studies

The foregoing findings prompted us to develop the three week treatment protocol to allow us to assess the effects of other antihypertensive agents on hemodynamics and structure. In these early reports we learned that some agents reduced cardiac mass after this very short-term treatment period of three weeks. However, in contrast, other anti-hypertensive agents, having perhaps even more salutary hemodynamic actions, did not reduce cardiac mass during the same treatment period (28,29).

The prototype of those antihypertensive agents that rapidly reduce cardiac mass was the centrally-acting adrenergic inhibiting agent methyldopa. Methyldopa reduced arterial pressure through a fall in total peripheral resistance in the SHR and of cardiac output in the WKY; and this was associated with a reduced cardiac mass (28,29). Reports by Sen and Tarazi indicated that this reduction in cardiac mass was related to reduced cardiac muscle protein and an increased proportion of collagen (33, 34), thereby suggesting a true reduction in myocardial muscle mass with an increased proportion of connective tissue in this so called "regressed hypertrophy heart". These findings were complicated by the observation that even the hearts of the normotensive WKY rats without LVH had a reduced cardiac mass and a decreased cardiac output. Furthermore, they underscored our earlier observation of a functional/structural dissociation.

And, further supporting this hypothesis were our findings with the dextroisomer of methyldopa and the responses to clonidine, another centrally active anti-adrenergic agent (29). However, in these studies cardiac mass was not reduced in SHR rats treated with d-methyldopa or clonidine; d-methyldopa had no hemodynamic effects even though clonidine produced the same hemodynamic effects in the SHR as methyldopa. Moreover, unlike methyldopa, clonidine failed to reduce cardiac mass in either the SHR or WKY (29). But, when the dose of clonidine was tripled, arterial pressure and total peripheral resistance increased by virtue of its agonistic action on peripheral post-synaptic alpha-adrenergic receptors; yet cardiac mass became reduced in the SHR, but not in the WKY. Thus, these effects in the WKY and SHR were very different from methyldopa and once again further supported the concept of a functional and structural dissociation of effects on the heart by certain pharmacological agents.

Using this very same protocol we treated WKY and SHR rats with the direct-acting smooth muscle vasodilating agent hydralazine (29). Despite a marked reduction of arterial pressure in both strains of rats (WKY and SHR) associated with a significant fall in total peripheral resistance, cardiac mass was not reduced. Moreover, in other studies we have also treated WKY and SHR rats with the post-synaptic alpha-adrenergic receptor inhibitors prazosin (35,36) and urapidil (36), the latter a single molecular agent that has both a central action like clonidine and a peripheral action like prazosin. Neither drug reduced cardiac mass despite their effective control of arterial pressure achieved through an inhibition of the adrenergic nervous system.

## Early Studies of Cardiac Performance

Our first studies that were designed to determine whether the once-hypertrophied left ventricle was able to improve its pumping ability were performed in rats with two-kidney, one-clip Goldblatt hypertension (2-K, 1-C, GH) (31). In these studies, 16 to 20 week old male WKY rats with 2-K, 1-C, GH for four weeks demonstrated an increased cardiac mass that was in direct proportion to the height of elevated arterial pressure; and their left ventricular pumping ability was significantly impaired. When their arterial pressure and total peripheral resistance were normalized pharmacologically (with methyldopa) or surgically (by removing the renal arterial clip), left ventricular mass was reduced to the normal weight of sham operated rats; and their left ventricular pumping ability was similarly normalized. However, these studies failed to answer the question as to whether improved left ventricular pumping ability would be maintained if arterial pressure were abruptly increased to the pretreatment hypertensive levels–just as the treated patient would demonstrate an abrupt pressure elevation when antihypertensive therapy was suddenly discontinued (37,38).

## Recent Studies

The foregoing studies prompted a further and more intensive exploration of the effects of the three classes of antihypertensive agents that consistently reduced cardiac mass in the SHR within three weeks of treatment: methyldopa (15,28,29,31); the angiotensin converting enzyme (ACE) inhibitors (16,17,39); and the calcium antagonists (18,40,41). In these studies all rats were treated for three weeks, and their effects on cardiovascular structure were assessed by measuring the mass of the left and right ventricles and a measured segment of the aorta. Function was assessed by measuring pumping ability of the left ventricle at the pharmacologically reduced pressures as well as after pressure was abruptly increased with the aortic snare and by determining aortic distensibility (as described above) (15-18,30,31) (see Tables 1-3).

## Methyldopa

As reported in our earlier studies, cardiac mass of both the WKY and the SHR was reduced with methyldopa; but when this was determined for both the left and right ventricles, our evidence of a structural/functional dissociation was further supported. Both ventricular masses were reduced; and this occurred in both rat strains (Table 1) (15); these surprising findings were complicated and confused further by the observation that neither the normal nor the thickened aortic mass were reduced in the WKY or the SHR, respectively. The structural cardiac changes were associated with reduced pumping ability in the WKY which became further impaired when left ventricular afterload was increased to pretreatment levels. However, responses in the SHR were very different: pumping ability with treatment was not diminished further unless the pressure and left ventricular was increased abruptly to pretreatment levels with occlusion of the aortic snare. The difference in these findings was explained by the additional observation that even though aortic mass was not reduced in either strain, *in vivo* aortic distensibility remained unchanged in the WKY but was reduced (i.e., improved) in the SHR (Table 3). Thus, in the SHR, methyldopa improved left ventricular impedance and reduced aortic compliance without reducing aortic mass; and this was not observed in the WKY even though both left and right ventricular masses were reduced.

# TABLE 1

## EFFECTS OF ANTIHYPERTENSIVE AGENTS ON CARDIOVASCULAR MASS

| Antihypertensive Agents | WKY Rats | | | | SHR Rats | | | |
|---|---|---|---|---|---|---|---|---|
| | Ventricular | | | Ascending Aorta | Ventricular | | | Ascending Aorta |
| | Total | LV | RV | | Total | LV | RV | |
| Methyldopa (15) | ↓ | ↓ | ↓ | NC | ↓ | ↓ | ↓ | NC |
| *ACE Inhibitors:* | | | | | | | | |
| Captopril (16) | NC | ↓ | NC | | ↓ | ↓ | ↓ | |
| CGS-16617 (17) | ↓ | ↓ | NC | NC | ↓ | ↓ | NC | ↓ |
| Cilazapril (17) | ↓ | ↓ | NC | ↓ | ↓ | ↓ | NC | ↓ |
| Quinapril (17) | NC | NC | NC | NC | ↓ | ↓ | NC | ↓ |
| *Calcium Antagonists:* | | | | | | | | |
| Nitrendipine (18) | NC | NC | ↑ | ↓ | ↓ | ↓ | NC | ↓ |
| Nifedipine (18) | NC | NC | ↑ | NC | ↓ | ↓ | NC | ↓ |
| Nisodipine (18) | ↑ | NC | ↑ | ↓ | ↓ | ↓ | ↑ | ↓ |
| Diltiazem (41) | NC | | | | NC | | | |

Summary of findings published from various studies from our laboratory (15-18,41). ↓ = reduced mass; NC = no change; ↑ = increased mass; a blank space reflects not reported.

## ACE Inhibitors

Unlike our findings with methyldopa, left ventricular mass was reduced in both the WKY and the SHR with captopril, but right ventricular mass was reduced only in the SHR (Table 1) (16). These structural alterations were associated with no changes in the pumping ability of the left ventricle at the pharmacologically reduced afterload. However, when arterial pressure was increased abruptly by tightening the aortic snare so that pressure rose to pretreatment levels, left ventricular pumping ability became impaired in the SHR but remained improved in the WKY (Table 2).

Similar studies were performed with three other ACE inhibitors, CGS-16617, cilazapril, and quinapril (17). The findings became still more intriguing since the changes within this same pharmacological group of antihypertensive agents were variable even though the hypertrophied SHR left ventricle demonstrated reduced mass with all three compounds (Table 1). Quinapril not only failed to reduce left ventricular mass in the WKY but it also failed to reduce right ventricular mass in both strains. This inability of quinapril to reduce right ventricular mass was also observed with CGS-16617 and cilazapril (Table 1). Furthermore, although aortic mass was reduced in the SHR with all three ACE inhibitors, only cilazapril reduced aortic mass in the WKY. Changes in cardiac pumping ability and aortic distensibility were similarly variable. Thus, none of these ACE inhibiting agents impaired left ventricular pumping ability in the SHR; both cilazapril and quinapril improved pumping ability in the SHR; although it remained unchanged with CGS-16617 (Table 2). When pressure was increased (by tightening the aortic snare) to pretreatment arterial pressures in the SHR, cilazapril further improved pumping performance, but performance remained unchanged with quinapril and became impaired with CGS-6617. The findings were still more disparate in the WKY: cilazapril improved pumping ability at both reduced and increased pressure loads; performance remained unchanged with quinapril at both lowered and increased loads; and with CGS-16617, performance was reduced at the pharmacologically reduced afterload but did not become impaired further with increased afterload. Aortic distensibility (*in vivo*) for the most part either remained unchanged or became slightly impaired (reduced) with all three agents, although CGS-16617 improved distensibility at the reduced afterload (Table 3). However, when aortic distensibility was assessed *in vitro*, it improved with all three agents in the WKY and with CGS-16617 and cilazapril in the SHR; it remained unchanged with quinapril (Table 3).

## Calcium Antagonists

The same variability in responses of cardiovascular structure and function within the same pharmacological class of antihypertensive agents in both SHR and WKY was again found with three different calcium antagonists (i.e., nitrendipine, nifedipine, and nisoldipine) (18). These studies demonstrated reduced left ventricular mass with all three agents in the SHR whereas WKY left ventricular mass remained unchanged with all three agents. Neither nitrendipine nor nifedipine changed right ventricular mass in the SHR. However, all three agents increased WKY right ventricular mass; and nisoldipine also increased right ventricular mass in the SHR (Table 1). These were the first findings of pharmacologically-induced isolated right ventricular enlargement with antihypertensive agents associated with either reduced or unchanged left ventricular mass. Moreover, these surprising findings produced by these

**TABLE 2**

**EFFECTS OF ANTIHYPERTENSIVE AGENTS ON LEFT VENTRICULAR PUMPING ABILITY**

| Antihypertensive Agents | WKY Rats | | SHR Rats | |
|---|---|---|---|---|
| | Reduced Load | Increased Load | Reduced Load | Increased Load |
| Methyldopa (15) | ↓ | ↓ | NC | ↓ |
| *ACE Inhibitors:* | | | | |
| Captopril (16) | NC | NC | NC | ↓ |
| CGS-16617 (17) | ↓ | NC | NC | ↓ |
| Cilazapril (17) | ↑ | ↑ | ↑ | ↑ |
| Quinapril (17) | NC | NC | ↑ | NC |
| *Calcium Antagonists (18):* | | | | |
| Nitrendipine | ↑ | ↑ | ↑ | NC |
| Nifedipine | ↑ | ↑ | ↑ | NC |
| Nisodipine | ↑ | ↑ | NC | ↓ |

Summary of findings published from various studies from our laboratory (15-18,41). ↓ = reduced pumping ability; NC = no change.

183

calcium antagonists were unassociated with any difference in their hemodynamic effects or those produced by the other antihypertensive agents. Providing still further evidence of differences within this class of agents was our earlier finding of unchanged left ventricular mass with diltiazem in either the WKY or SHR despite its similar hemodynamic effects as with the other calcium antagonists (41). Nevertheless, all three calcium antagonists improved WKY left ventricular pumping ability at both the reduced as well as increased left ventricular afterloads, although the SHR findings were more variable (Table 2). SHR pumping ability was improved with both nitrendipine and nifedipine at the reduced afterload; but it remained unchanged (with respect to pretreatment pumping ability) with nisoldipine, although it became improved when the afterload was increased (Table 2). Finally *in vivo* WKY aortic distensibility became impaired with nitrendipine and nisoldipine but remained unchanged with nifedipine at the pharmacologically reduced pressures. However, *in vivo* aortic distensibility decreased in all three WKY and SHR rat calcium antagonist groups at the increased afterload. In contrast, at the pharmacologically reduced pressures in the SHR, *in vivo* aortic distensibility improved with nitrendipine and nifedipine but diminished with nisoldipine (Table 3). More consistent however, were the findings of *in vitro* aortic distensibility which improved with all three calcium antagonists in both the WKY and SHR rats.

DISCUSSION

These data, derived from normotensive and hypertensive rats treated with agents from three pharmacological classes of drugs known to reduce cardiac mass within a very short time period, demonstrated several major findings. First, to assume that these agents only reduce the mass of hypertrophied ventricles or vessels is incorrect. Some of these agents reduced the mass of non-hypertrophied ventricles; and this was independent of their effects on aortic mass (e.g., methyldopa). The effects of agents within the same seemingly homogeneous class of agents (i.e., ACE inhibitors) were likewise variable. All four ACE inhibitors that were studied reduced SHR left ventricular mass; but, in the WKY, only quinapril failed to reduce left ventricular mass. In contrast to methyldopa, all ACE inhibitors (except captopril in the SHR) were unable to reduce right ventricular mass; and, whereas the ACE inhibitors reduced aortic mass in the SHR, only cilazapril reduced WKY mass. And, most perplexing, while the calcium antagonists had either no effect (WKY) or reduced (SHR) left ventricular mass, all three agents increased right ventricular mass in the WKY, and nisoldipine increased right ventricular mass in the SHR.

Not only were these changes in cardiovascular mass highly variable but so were the associated performance (left ventricular pumping ability or aortic distensibility) studies variable. In general, pumping ability was diminished in both WKY and SHR with methyldopa, but remained unchanged, improved or diminished with the ACE inhibitors, and improved (WKY) or remained unchanged or improved (SHR) with the calcium antagonists. These changes could not be related to changes in aortic mass or distensibility, although *in vitro* distensibility tended to improve or remain unchanged with all agents (whether or not mass diminished).

**TABLE 3**

**EFFECTS OF ANTIHYPERTENSIVE AGENTS ON AORTIC DISTENSIBILITY**

| Antihypertensive Agents | WKY Rats | | | SHR Rats | | |
|---|---|---|---|---|---|---|
| | In Vivo Loading | | In Vitro | In Vivo Loading | | In Vitro |
| | Reduced | Increased | | Reduced | Increased | |
| Methyldopa (15) | NC | ↓ | | ↑ | ↓ | |
| *ACE Inhibitors:* | | | | | | |
| CGS-16617 (17) | ↓ | NC | ↑ | ↑ | ↓ | ↑ |
| Cilazapril (17) | ↓ | NC | ↑ | NC | ↓ | ↑ |
| Quinapril (17) | NC | ↓ | ↑ | NC | ↓ | NC |
| *Calcium Antagonists (18):* | | | | | | |
| Nitrendipine | ↓ | ↓ | ↑ | ↑ | ↓ | ↑ |
| Nifedipine | NC | ↓ | ↑ | ↑ | ↓ | ↑ |
| Nisodipine | ↓ | ↓ | ↑ | ↑ | ↓ | ↑ |

Summary of findings published from various studies from our laboratory (15-18,41). ↓ = reduced distensibility; NC = no change; ↑ = increased or improved distensibility.

As indicated at the onset of this discussion, an independent risk of increased cardiovascular morbidity and mortality has been ascribed to left ventricular hypertrophy. This assertion seems incontrovertible; but to suggest that by pharmacologically reducing left ventricular mass (i.e., LVH) the risk is improved has yet to be demonstrated. Even if risk is improved with therapy that reduced left ventricular mass, it would be difficult to relate this solely to the reversal of LVH *per se*. One would then question whether this improvement was also related to: (1) the reduced cardiac mass itself; (2) the associated arterial pressure; (3) the improved coronary arterial blood flow (either by absolute supply to the heart, increased delivery per gram of myocardial tissue, or increased coronary flow reserve); (4) the antiarrhythmic effects of the administered antihypertensive drug itself, or (5) those independent molecular/cellular effects of the antihypertensive drug on the cardiac myocytes or fibroblast.

With these concepts in mind, it would seem that at this time, it is premature to assume that reduction of left ventricular mass with antihypertensive therapy produces an associated reduction in risk. Thus, if evaluation of the patient with reduced cardiac mass is made while the patient is still receiving antihypertensive drugs (or even has received such therapy within the past several weeks), the improvement in myocardial performance or remission of cardiac dysrhythmias may only reflect the effect on pressure reduction, on cardiac rhythm, or on myocardial perfusion. It was for these reasons that the left ventricular pumping ability studies were performed at the reduced as well as elevated pressure levels.

Moreover, since the variety of drug studies produced disparate effects on cardiovascular structure and function, even within the same pharmacological class of agents, it is important to understand the underlying mechanism(s) that could explain these observations. It is clear that the hemodynamic alterations that relate to the development and reversal of ventricular hypertrophy are very important. This role is not denied; however, non-hemodynamic factors must also be considered. Among these factors several may be excluded–differences that may be ascribed to the rat strain, age, or gender. Homogeneous strains of WKY and SHR rats were used; and they were of the same age and gender. Moreover, the same investigators conducted the studies using the same protocol.

Thus, we are left with differences among the drugs (even within the same class of agents) that relate to their peculiar pharmacokinetics and pharmacodynamic actions. Yet these differences do not explain why certain agents (i.e., calcium antagonists) have the capability of increasing right ventricular mass while they reduce left ventricular mass. In this regard, it is possible that intramyocytic concentration of the calcium ion may play an important role in protein synthesis.

The studies involving the ACE inhibitors although qualitatively and quantitatively different from those of the calcium antagonists and methyldopa, suggest the important autocrine (paracrine and even intracrine) role of the intramyocytic renin-angiotensin systems; and several reports have already incriminated the role of this system on protein synthesis (42).

The common denominator for the dissociation of cardiovascular structural and functional effects of the many antihypertensive drugs may reside in the interaction of these agents on natural intracellular protooncogenes that influence protein synthesis. Studies have already demonstrated the interaction of adrenergic stimulation of the cardiac myocyte on *c-myc* mediated DNA synthesis (43). Other studies have implicated other proto-oncogenes (e.g., *c-fos*, *c-ras*, *c-mas*) intracellular events mediated through the renin-angiotensin system (5,44). Whether the different naturally occurring humoral or intracellular systems or messengers have specific effects on their own respective intracellular growth factors, or whether stimulation of inhibition of these factors are dependent upon optimal drug penetrance and action is unknown at this time. A number of laboratories world-wide are actively exploring these very provocative and intriguing mechanisms; and, no doubt, important answers will be forthcoming in the very near future.

However, at this time it is premature to arrive at simplistic generalizations and inopportune clinical assertions. The important concept that must be remembered is that no matter what agent is used to treat the patient with hypertension, optimal long-term control of pressure should be the goal of therapy. It is this achievable therapeutic goal that has already effected the dramatic improvements in cardiovascular morbidity and mortality associated with hypertensive diseases.

## SUMMARY

The heart is one of the major target organs that becomes secondarily involved with the unrelenting and progressive vascular disease of essential hypertension. As a result of this increasing afterload that is imposed upon the left ventricle, the ventricular chamber adapts structurally and functionally. Structural changes involve an increase in muscle mass that is achieved through left ventricular hypertrophy (in a manner similar to the arteriolar changes demonstrated by increased thickening). Unless antihypertensive therapy is interdicted in this disease process, left ventricular failure will ensue as the major cardiac hemodynamic consequence. Left ventricular hypertrophy is also associated with a risk that is independent of the pressure overload and hemodynamic risk. Although antihypertensive therapy will reduce from the hemodynamic alterations, only recently have epidemiological findings suggested that the independent risk of LVH may be reduced with pharmacological therapy. There are no data available to indicate just which agents may reduce the risk from LVH; but relatively recent studies seem to indicate that while all agents may reduce LVH with prolonged therapy only certain classes of agents will do so independent of their hemodynamic factors. Some of these agents, however, may impair cardiac function if arterial pressure is increased abruptly following therapeutic reduction of cardiac mass. Other agents may preserve normal function--or even may improve function.

Among those classes of antihypertensive agents that reduce cardiac mass at least in part due to nonhemodynamic factors, are the angiotensin converting enzyme inhibitors, the calcium antagonists, and most adrenergic inhibitors. Evidence will be presented demonstrating the hemodynamic/structural dissociation of those pharmacological agents that reduce cardiac mass with short-term treatment in spontaneously hypertensive rats with left ventricular hypertrophy. Although centrally active adrenolytic, angiotensin converting enzyme (ACE) inhibitors, and calcium antagonists all reduce cardiac mass, their structural and cardiac functional effects differ greatly. Even within the ACE inhibitor group their effects vary--improving, impairing, or not changing the Frank-Starling relationships following reduction in left ventricular mass. We postulate great variability of cardiac intramyocytic penetrance of the pharmacological agents and their local intracellular effects on mitogenesis of the ventricular myocyte. The implications on cardiac function and therapy have vast potential.

Therefore, current investigative areas involving new concepts of molecular biology of the cardiac myocyte may provide great promise to the quest of unraveling some of the newly postulated questions: What is the role of ionized intracellular calcium? Do the local renin-angiotensin systems in the cardiac and vascular myocyte participate in the development and regression of hypertrophy? Do they involve, more specifically, certain cellular mitogens such as the proto-oncogenes or local intracellular growth factors? And, of course, will therapeutic "regression" of left ventricular hypertrophy be specifically associated with reduced risk? And, finally, by learning the specific cellular biological "switch" that "shuts off" protein synthesis, will it be possible to synthesize an agent that will conversely stimulate protein synthesis and thereby offer promise of a new and "natural" inotropic agent for patients with cardiac failure and myocardiopathies?

# REFERENCES

1.  Frohlich ED. The heart in hypertension. *In: Hypertension: Physiopathology and Treatment,* 2nd ed, J. Genest, O. Kuchel, P. Hamet, M. Cantin (eds). New York: McGraw-Hill, pp 791-810, 1983.
2.  Frohlich ED. Hemodynamics and other determinants in development of left ventricular hypertrophy: Conflicting factors in its regression. *Fed Proc* 42: 2709, 1983.
3.  Frohlich ED. Cardiac hypertrophy: Stimuli and mechanisms. *In: Scientific Foundations of Cardiology*, P. Sleight (ed). London: Heinemann, pp 182-190, 1983.
4.  Frohlich ED. The heart in hypertension: Unresolved conceptual challenges. *Hypertension* 11: 19, 1988.
5.  Frohlich ED. The First Irvine H. Page Lecture (State of the Art). The mosaic of hypertension: Past, present, and future. *J Hypertension* 6: 42, 1989.
6.  Frohlich ED. The heart in hypertension. *In: Arterial Hypertension*, 2nd ed., J. Rosenthal, A.V. Chobanian (eds). Berlin: Springer-Verlag, (in press).
7.  Folkow B, Hallback M, Lundgren Y, Sivertsson R, Weiss L. Importance of adaptive changes in vascular design for establishment of primary hypertension studied in man and in spontaneously hypertensive rats. *Circ Res* 32: 2, 1973.
8.  Kannel WB, Gordon T, Offutt D. Left ventricular hypertrophy by electrocardiogram: Prevalence, incidence and mortality in the Framingham study. *Ann Intern Med* 71: 89, 1969.
9.  Messerli FH, Ventura HO, Elizardi DJ, Dunn FG, Frohlich ED. Hypertension and sudden death: Increased ventricular ectopic activity in left ventricular hypertrophy. *Am J Med* 77: 18, 1984.
10. Frohlich ED. Left ventricular hypertrophy as a risk factor. *In: Cardiology Clinics*, Vol. 4, F.H. Messerli, C. Amodeo (eds). Philadelphia: W.B. Saunders, pp 137-144, 1986.
11. Kannel WB, D,Agostino RB, Levy D, Belanger AJ. Prognostic significance of regression of left ventricular hypertrophy (abstract). *Circulation* 78(Suppl II): II-89, 1988.

12. Frohlich E D. Cardiac hypertrophy in hypertension (editorial). *N Engl J Med* 317: 831, 1987.
13. Frohlich ED, Pfeffer MA, Pfeffer JM. Systemic hemodynamics and cardiac function in spontaneously hypertensive rats: Similarities with essential hypertension. *In: The Heart in Hypertension,* B.E. Straub (ed). Berlin: Springer-Verlag, pp 425-435, 1981.
14. Frohlich ED. Effet du traitement antihypertenseur sur l'hypertrophie ventriculaire qauche dans l'hypertension afterielle essentielle. (Neckers Seminars-Paris) *Flammarion Medecine-Sciences,* pp 87-100, 1989.
15. Sasaki 0, Kardon MB, Pegram BL, Frohlich ED. Aortic distensibility and left ventricular pumping ability after methyldopa in Wistar-Kyoto and spontaneously hypertensive rats. *J Vasc Med Biol* 1: 59, 1981.
16. Natsume T, Kardon MB, Pegram BL, Frohlich ED. Ventricular performance in spontaneously hypertensive rats with reduced cardiac mass. *Cardiovasc Drugs Ther* 3: 433, 1989.
17. Frohlich ED, Sasaki 0. Dissociation of changes in cardiovascular mass and performance with angiotensin converting enzyme inhibitors in Wistar-Kyoto and spontaneously hypertensive rats. *J Am Coll Cardiol* (in press).
18. Frohlich ED, Sasaki 0. Calcium antagonists variably change cardiovascular mass and improve function in rats (abstract). *J Am Coll Cardiol* 15: 110A, 1990.
19. Frohlich ED, Tarazi RC. Is arterial pressure the sole factor responsible for hypertensive cardiac hypertrophy? *Am J Cardiol* 44: 959, 1979.
20. Tarazi RC, Frohlich ED. Is reversal of cardiac hypertrophy a desirable goal of antihypertensive therapy? *Circulation* 75: 113, 1987.
21. Frohlich ED, Messerli FH, Re RN, Dunn FG. Mechanisms controlling arterial pressure. *In: Pathophysiology: Altered Regulatory Mechanisms in Disease,* 3rd ed, E.D. Frohlich (ed). Philadelphia: J.B. Lippincott, pp 45-81, 1984.
22. Trippodo NC, Frohlich ED. Controversies in cardiovascular research: Similarities of genetic (spontaneous) hypertension. Man and rat. *Circ Res* 48: 309, 1981.
23. Frohlich ED. Is the spontaneously hypertensive rat a model for human hypertension? *J Hypertension* 4(Suppl 3): 15, 1986.
24. Frohlich ED, Pfeffer MA, Pfeffer JM. Systemic hemodynamics and cardiac function in spontaneously hypertensive rats: similarities with essential hypertension. *In: The Heart in Hypertension,* B.E. Straub (ed). Berlin: Springer-Verlag, pp 425-435, 1981.
25. Pfeffer MA, Pfeffer JM, Dunn FG, Nishiyama K, Tsuchiya M, Frohlich ED. Natural ventricular hypertrophy in normotensive rats. I. Physical and hemodynamic characteristics. *Am J Physiol* 236: H640, 1979.
26. Sesoko S, Pegram BL, Kuwajima I, Frohlich ED. Hemodynamic studies in spontaneously hypertensive rats with congenital arteriovenous shunts. *Am J Physiol* 242: H722, 1982.
27. Pfeffer JM, Pfeffer MA, Frohlich ED. Validity of an indirect tail-cuff method for determining systolic arterial pressure in unanesthetized normotensive and spontaneously hypertensive rats. *J Lab Clin Med* 78: 957, 1971.
28. Ishise S, Pegram BL, Frohlich ED. Disparate effects of methyldopa and clonidine on cardiac mass and haemodynamics in rats. *Clin Sci* 59(VI): 449s, 1980.

29. Pegram BL, Ishise S, Frohlich ED. Effect of methyldopa, clonidine, and hydralazine on cardiac mass and haemodynamics in Wistar-Kyoto and spontaneously hypertensive rats. *Cardiovasc Res* 16: 40, 1982.

30. Pfeffer MA, Pfeffer JM, Frohlich ED. Pumping ability of the hypertrophying left ventricle of the spontaneously hypertensive rat. *Circ Res* 38: 423, 1976.

31. Kuwajima I, Kardon MB, Pegram BL, Sesoko S, Frohlich ED. Regression of left ventricular hypertrophy in two-kidney, one clip Goldblatt hypertension. *Hypertension* 4:113, 1982.

32. Pfeffer MA, Pfeffer JM, Weiss AK, Frohlich ED. Development of SHR hypertension and cardiac hypertrophy during prolonged beta blockade. *Am J Physiol* 232: H639, 1977.

33. Sen S, Bumpus FM. Collagen synthesis in development and reversal of cardia hypertrophy in spontaneously hypertensive rats. *Am J Cardiol* 44: 954, 1979.

34. Sen S, Tarazi RC, Bumpus FM. Biochemical changes associated with development and reversal of cardiac hypertrophy in spontaneously hypertensive rats. *Cardiovasc Res* 10: 254, 1976.

35. Pegram BL, Frohlich ED. Cardiovascular adjustment to anti-adrenergic agents. *Am J Med* 75: 94, 1983.

36. Pegram BL, Kobrin I, Natsume T, Gallo AJ, Frohlich ED. Systemic and regional hemodynamic effects of acute and prolonged treatment with urapidil or prazosin in normotensive and spontaneously hypertensive rats. *Am J Med* 77: 64, 1984.

37. Dustan HP, Page IH, Tarazi RC, Frohlich ED. Arterial pressure responses to discontinuing antihypertensive drug treatment. *Circulation* 37: 370, 1968.

38. Frohlich ED. Clinical conference: Hypertensive cardiovascular disease. A pathophysiological assessment. *Hypertension* 6: 934, 1984.

39. Pegram BL, Frohlich ED. Immediate systemic and regional hemodynamic effects of MK-421 converting enzyme inhibitor in conscious Wistar-Kyoto and spontaneously hypertensive rats. *In: Hypertensive Mechanisms of the Spontaneously Hypertensive Rat as a Model to Study Human Hypertension*, W. Rascher, D. Clough, D. Ganten (eds). New York: Schattauer FK, p 677, 1982.

40. Kobrin I, Sesoko S, Pegram BL, Frohlich ED. Reduced cardiac mass by nitrendipine is dissociated from systemic or regional haemo-dynamic changes in rats. *Cardiovasc Res* 3: 158, 1984.

41. Natsume T, Gallo A, Pegram BL, Frohlich ED. Hemodynamic effects of prolonged treatment with diltiazem in conscious normo-tensive and spontaneously hypertensive rats. *Clin Exper Hyper* A7: 1471, 1985.

42. Frohlich ED, Iwata T, Sasaki 0. Clinical and physiologic signifi-cance of local tissue renin-angiotensin systems. *Am J Med* 87: 19, 1989.

43. Starksen NF, Simpson PC, Bishopric N, Coughlin SR, Lee WMF, Escobedo JA, Williams LT. Cardiac myocyte hypertrophy is associ-ated with c-myc protooncogene expression. *Proc Natl Acad Sci* 93: 9349, 1986.

44. Jackson TR, Blair AC, Marshall M, Geodert M, Hanley MR. The mas oncogene encodes an angiotensin receptor. *Nature* 335: 437, 1988.

# Ca/CaM-STIMULATED AND cGMP-SPECIFIC PHOSPHODIESTERASES

# IN VASCULAR AND NON-VASCULAR TISSUES

H.S. Ahn, M. Foster, M. Cable, B.J.R. Pitts and E.J. Sybertz

Department of Pharmacology
Schering-Plough Research Division
Bloomfield, NJ 07003

## INTRODUCTION

A large body of evidence indicates that guanosine 3',5'-cyclic monophosphate (cGMP) plays an essential role in vasorelaxant actions of various agents such as atrial natriuretic factor (ANF), nitrogen oxide containing compounds (e.g., nitroprusside) and endothelium dependent vasodilators (e.g., acetylcholine) (1). These agents elevate cGMP by stimulating either soluble or particulate guanylate cyclase (1).

In addition to guanylate cyclase, cyclic nucleotide phosphodiesterase (PDE) appears to play an important role in regulating cGMP levels and thus vasorelaxation. Although many PDEs have been identified (2), vascular tissues are known to contain two to three forms of PDE– 1) calcium/calmodulin-stimulated PDE (CaM-PDE or type I PDE), 2) cGMP-specific PDE (cG-PDE or type III), and 3) cAMP-PDE or type IV PDE (4-6).

Recently, selective PDE inhibitors became available and have been used to characterize different forms of PDEs. In vascular tissues, vinpocetine and 8-methoxymethyl isobutylmethylxanthine (8-MeOMeMIX) were reported to be selective inhibitors of CaM-PDE (3,7), and dipyridamole and M&B 22948 selective cG-PDE inhibitors (5,6,8).

Although both CaM-PDE and cG-PDE were present and accounted for most cGMP hydrolysis in aorta (9), the relative importance of these PDEs in cGMP hydrolysis is not known. Also, it is not clear whether or not vascular tissue CaM-PDE differs from non-vascular CaM-PDE. The purpose of this study was to determine the relative activity of CaM-PDE and cG-PDE in various tissues and cultured cells using selective inhibitors and a conformation-specific anti-calmodulin monoclonal antibody (10) which selectively binds CaM-PDE. In addition, CaM-PDE was purified from porcine aortic extracts and was compared to a similarly purified bovine brain enzyme. Results of this study show similar properties of aortic and brain CaM-PDEs but a large difference in relative activity of CaM-PDE and cG-PDE in various tissues and cell types.

*Cellular and Molecular Mechanisms in Hypertension*
Edited by R.H. Cox, Plenum Press, New York, 1991

# METHODS

## PDE Assays

PDE activity was measured by a radioenzymatic assay as previously described (11). Enzyme or tissue extract was incubated with 1-2 $\mu$M of unlabelled substrate and tritiated substrate (about 60,000 cpm of $^3$H-cGMP or $^3$H-cAMP) in 50 mM Tris-HCl buffer, pH 7.5, containing 2 mM $MgCl_2$, 0.1 $\mu$M CaM and 0.2 mM $CaCl_2$ at 30°C for 15 min.

## Culture of Bovine Aortic Endothelial Cells and Rat Aortic Smooth Muscle Cells

Both cell cultures were propagated from frozen cells at early (2 to 4) passages. Bovine aortic endothelial cells were cultivated in RPMI 1640 medium (Gibco, NY) supplemented with 20% fetal calf serum (Hyclone, UT), penicillin (100 units), streptomycin (100 $\mu$g/ml) and 2 mM of L-glutamine at 37°C in a humidified atmosphere of 95% air-5% $CO_2$. Rat aortic smooth muscle cells were cultured in Dulbecco's modified essential medium supplemented with 10% fetal calf serum and penicillin-streptomycin as above.

## Preparation of Cell and Tissue Extracts

Confluent bovine aortic endothelial or rat aortic smooth muscle cells were washed once with Dulbecco's phosphate buffered saline and scraped in 2 ml of a homogenization buffer from 3-100 mm plates. The harvested cells were homogenized using an all glass Kontes dual homogenizer. The homogenization buffer was 50 mM Tris-HCl, pH 7.5, containing 5 mM $MgCl_2$, 0.1 mM of phenylmethylsulfonyl fluoride (PMSF), 10 $\mu$M leupeptin and 0.2 mg/ml BSA. Various porcine tissues were homogenized in 3 to 10 volume (v/w) of the homogenization buffer. The cell and tissue homo-genates were centrifuged at 2000 g for 20 min at 5°C. These supernatants were used in immunoadsorption assays.

## Immunoadsorption of CaM-PDE in Isolated CaM-PDE Preparation, Tissue and Cultured Cell Extracts

A solid phase CaM-PDE monoclonal antibody (ACC-1 linked to Pansorbin cell) was prepared and immunoadsorption assays carried out as previously described (11). Aliquots (200-300 $\mu$l) of an isolated CaM-PDE or the 2000 g supernatant of tissue and cell extracts were incubated with the native or boiled solid phase ACC-1 (1 $\mu$g) or C1 in the presence of CaM (0.5 $\mu$g/0.2 mL) and $CaCl_2$ (0.1 mM) at 5°C for 2 to 3 hr, and were centrifuged at 10,000 g. The supernatants and resuspended pellets were assayed for PDE activity using 2 $\mu$M cGMP as substrate.

## Purification of CaM-PDE

CaM-PDE was purified from porcine aorta media layer by successive chromatography on DEAE-sephacel column and CaM affigel 15 column. The minced porcine aortic media was homogenized in 3 vol (v/w) of 50 mM Tris-acetate buffer (pH 6.0) containing 3.75 mM 2-mercaptoethanol, 0.1 mM PMSF and 10 $\mu$M leupeptin (PDE buffer) using a Brinkman PT-10 polytron (2 x 15 sec bursts at the maximal speed). The homogenate was centrifuged at 100,000 g for 60 min. The resulting supernatant was applied to a DEAE-sephacel column equilibrated with the PDE isolation buffer. PDE peaks

**TABLE 1**

**IMMUNOADSORPTION OF cGMP-HYDROLYZING ACTIVITY
BY SOLID PHASE CaM MONOCLONAL ANTIBODY IN
EXTRACTS OF VARIOUS TISSUES AND CELLS**

| Tissue or Cell | Conditions of Antibody | Supernatant | Pellet | |
|---|---|---|---|---|
| | | cGMP-hydrolyzing activity (pmol/tube/min) | | |
| Pig aorta | boiled | $13.60 \pm 0.03$ | $0.46 \pm 0.07$ | |
| | native | $3.92 \pm 0.17$[a] | $11.20 \pm 0.23$[b] | (73%) |
| Pig coronary artery | boiled | $7.80 \pm 0.22$ | $0.10 \pm 0.0$ | |
| | native | $2.20 \pm 0.22$ | $6.80 \pm 0.15$ | (75%) |
| Pig lung | boiled | $12.10 \pm 0.05$ | $0.29 \pm 0.01$ | |
| | native | $11.70 \pm 0.22$ | $3.65 \pm 0.12$ | (22%) |
| Pig heart | boiled | $5.03 \pm 0.05$ | $0.17 \pm 0.05$ | |
| | native | $3.37 \pm 0.74$ | $2.16 \pm 0.06$ | (37%) |
| Pig brain | boiled | $3.28 \pm 0.12$ | $0 \pm 0.06$ | |
| | native | $1.01 \pm 0.08$ | $1.50 \pm 0.03$ | (60%) |
| Bovine aortic endothelial cells[c] | boiled | 87 | 5.3 | |
| | native | 76 | 9.5 | (5%) |
| Rat aortic smooth muscle cells[c] | boiled | 75 | 6.1 | |
| | native | 38 | 71 | (63%) |

Each value represents an average of 2 determinations or the mean ± SEM of 3 to 4 determinations. Percent of total cGMP-hydrolyzing activity which was immuno-adsorbed is in parenthesis. Confluent endothelial cells (passage 5) and smooth muscle cells (passage 4) were used as enzyme sources. [a]Not inhibited by 1 mM EGTA but inhibited by dipyridamole (IC$_{50}$, 8 µM); [b]Inhibited by 1 mM EGTA; [c]pmol cGMP hydrolyzed/min/100 mm plate of cultured cells.

TABLE 2

**PURIFICATION OF PIG AORTIC CaM-PDE**

| | Vol (ml) | Protein (mg) | Total Activity Ca/CaM (nmol/min) | EGTA (nmol/min) | Specific Activity Ca/CaM (nmol/mg/min) |
|---|---|---|---|---|---|
| Homogenate | 1,500 | 36,900 | 2,945 | 1,749 | 0.08 |
| 100 K Supernatant | 1,352 | 26,499 | 1,494 | 1,455 | 0.06 |
| DEAE-sephacel | 90 | 251 | 193 | 80 | 0.77 |
| CaM-affigel | 40 | 5.1 | 72 | 22 | 14 |

The starting material was 300 g of pig aortic media. Each fraction was assayed for cGMP-hydrolyzing activity in the presence of Ca (1 mM)/CaM (0.15 μM) or EGTA (1 mM) using 1 μM cGMP.

**TABLE 3**

**PROPERTIES OF VASCULAR AND BRAIN CaM-PDE ISOZYMES**

| Enzyme Source | $K_m$ (μM) cG | cA | $V_{max}$[a] cG | cA | pH optimum | Stimulation by CaM | Binding to ACC-1[b] | C1[c] |
|---|---|---|---|---|---|---|---|---|
| Porcine aorta | 2.8 | 12 | 177 | 41 | 8.0 | 2-3 fold | 100% | yes[d] |
| Bovine brain | 1.6 | 4.5 | 1531 | 710 | 8.0 | 3-5 fold | 95% | yes[d] |

Except for kinetic assays, assays were performed with 1 μM of cGMP or cAMP in the presence of CaM (0.5 μM)/CaCl$_2$ (0.1 mM) at 30°C for 5-15 min. The immunoadsorption experiments were carried out by incubating each enzyme with 0.1 ml of Pansorbin cells liked to ACC-1 (1 μg) for 2 hr or 0.1 ml of packed sepharose 4B linked to C1 monoclonal antibody for 30 min at 5°C. [a]nmol cyclic nucleotide hydrolyzed/mg protein/min; [b]monoclonal antibody to CaM bound to brain CaM-PDE (10); [c]monoclonal antibody to bovine brain 60 KDa CaM-PDE (14); [d]100 μl of packed sepharose linked to C1 specifically immunoadsorbed aortic CaM-PDE (0.41 nmol/min) and brain CaM-PDE (0.13 nmol/min).

were eluted using a step gradient as previously reported (12). The first peak containing CaM-stimulated activity was eluted with PDE isolation buffer containing 350 mM sodium acetate. The peak fractions were pooled, concentrated, dialyzed and was adjusted to 1 mM CaCl$_2$ and 100 mM NaCl and immediately applied to a CaM affigel column (1 ml bed vol./2.5 mg protein). The column was washed with the same buffer. CaM-PDE was eluted with 2 column volumes of PDE buffer containing 4 mM EGTA and 100 mM NaCl, concentrated and stored in 0.2% BSA at -70°C.

## RESULTS

Table 1 summarizes results of imunoadsorption studies with various tissue and cell extracts. The percent of cGMP-hydrolyzing activity immunoadsorbed onto solid phase ACC-1 varied widely from one tissue or cell to another. The immunoadsorbed activity was stimulated by Ca/CaM and represented CaM-PDE. The unadsorbed activity appeared to be mainly cG-PDE based on its selective inhibition by dipyridamole and M&B 22948, selective inhibitors of cG-PDE (data not shown). Thus, activities in the pellet and supernatant indicate CaM-PDE and cG-PDE activities respectively. For porcine tissues the highest immunoadsorbed CaM-PDE activity was found in coronary artery and aorta, with an intermediate activity in heart and the least activity in lung. In contrast, the immunoadsorbed CaM-PDE contributed less than 20% of total cAMP hydrolysis in all tissues examined (aorta, heart, lung and brain). These data indicate that cAMP is mainly hydrolyzed by other enzymes, probably low $K_m$ cAMP-PDE.

Sixty-three percent and 5% of the total cGMP-hydrolyzing activity were immunoadsorbed onto the monoclonal antibody in smooth muscle and endothelial cell extracts, respectively. The remaining activity in the supernatant was not stimulated by Ca/CaM and was inhibited by dipyridamole with respective $IC_{50}$'s of 4 to 8 μM. These data indicate that endothelial cells contain little CaM-PDE while smooth muscle cells contain both CaM-PDE and cG-PDE.

Table 2 shows purification of CaM-PDE from porcine aorta media. About 200-fold purification could be achieved by the successive ion exchange and CaM affinity chromatography. CaM-PDE was similarly purified about 300-fold also from bovine brain (data not shown). Table 3 summarizes results of studies comparing properties of porcine aortic and bovine brain CaM-PDEs. $K_m$ values of aortic and brain CaM-PDEs for cGMP and cAMP were comparable, but $V_{max}$ values of aortic CaM-PDE was 1/9 to 1/17 that of brain enzyme. CaM-stimulated activity of aortic homogenate was also 1/5 that of brain. Thus, these data suggest a lower concentration of CaM-PDE in aorta than in brain. Both CaM-PDEs showed similar properties in pH optimum, CaM-stimulability and cross-reactivity with monoclonal antibodies C-1 and ACC1 to brain CaM-PDE or CaM bound to brain CaM-PDE.

DISCUSSION AND CONCLUSION

In this study, CaM-PDE was separated from the rest of PDEs using a monoclonal antibody (ACC-1) selective for CaM-PDE, and cG-PDE was identified by selective inhibitors of cG-PDE after the removal of CaM-PDE by immunoadsorption in the tissue or cell extracts (10,13). The relative importance of CaM-PDE or cG-PDE was assessed by determining respective contribution of these two PDEs to the total cGMP hydrolysis in the extracts. Results indicate that, in addition to aorta, coronary artery and non-vascular tissues contained both CaM-PDE and cG-PDE, which accounted for most of cGMP hydrolysis. An exception was endothelial cells, which contained cG-PDE but little CaM-PDE. The relative contribution of CaM-PDE and cG-PDE to cGMP hydrolysis varied markedly among different tissues. For example, in the coronary artery and aorta, CaM-PDE hydro-lyzed a large portion of cGMP. In lung, cG-PDE appeared to hydrolyze most cGMP. This finding suggests a possibility that a selective CaM-PDE inhibitor may exhibit some degree of tissue selectivity in raising cGMP. CaM-PDE contributed minimally to cAMP hydrolysis in tissues examined. Since both cGMP-inhibited cAMP-PDE and rolipram-sensitive cAMP-PDE are known to occur in the aorta and other tissues (2,11), cAMP may be mainly hydrolyzed by these cAMP-PDEs in these tissues. Our data indicate the presence of cG-PDE activity in porcine coronary artery, where only CaM-PDE and low $K_m$ cAMP-PDE were previously detected (3).

The present study also compared properties of CaM-PDEs isolated from a vascular tissue (aorta) and a non-vascular tissue (brain). Although CaM-PDE was purified to homogeneity from several non-vascular tissues (2), vascular CaM-PDE has not been purified beyond an ion exchange chromatographic separation step (5-7). Our earlier immunoadsorption study indicated that a DEAE-sephacel column-purified rabbit aortic CaM-PDE preparation contained 28% of non-CaM-PDE activity (11). In contrast, the CaM-affinity column-purified porcine aortic CaM-PDE appears to contain no contaminating PDEs since its entire activity could be immuno-adsorbed by ACC-1 monoclonal antibody. These findings point out the desirability of using CaM-affinity purified CaM-PDE. Comparison of similarly purified aortic and brain CaM-PDEs shows that aortic and brain CaM-PDEs have similar kinetic properties, pH optimum and induce a similar conformation of CaM bound to them. These results extend our earlier studies showing a minimal difference in drug sensitivity of brain and aortic CaM-PDEs (6). Brain contained 60k Da and 63k Da subunits of CaM-PDE (14). C1 monoclonal antibody selectively interacted with the 60k Da subunit (14). Effective immunoadsorption of brain and aortic CaM-PDEs to C1 antibody in this study indicates that aortic CaM-PDE is related to the 60k Da subunit. Lung and heart CaM-PDE isozymes were reported to be also related to 60k Da but not 63k Da subunit (14).

In conclusion, aortic smooth muscle, vascular and non-vascular tissues contain CaM-PDE and cG-PDE in varying ratios. Endothelial cells lack CaM-PDE. In addition, this study combined with an earlier finding (6) indicates a similar property of vascular and non-vascular CaM-PDE isozymes in terms of kinetic behavior, antigenic structure and drug sensitivity.


## ACKNOWLEDGMENTS

The authors wish to thank Drs. J. Beavo and J. Wang for kindly providing ACC-1 and C1 monoclonal antibodies.

## REFERENCES

1.  Murad F.  Cyclic guanosine monophosphate as a mediator of vasodilation. J Clin Invest 78: 1-5, 1986.
2.  Beavo JA.  Multiple isozymes of cyclic nucleotide phosphodiesterase. *Adv Second Messenger Phosphoprotein Res* 22: 1-38, 1988.
3.  Lorenz KL, Wells JN.  Potentiation of the effects of sodium nitroprusside and of isoproterenol by selective phosphodiesterase inhibitors. *Mol Pharmacol* 23: 424-430, 1983.
4   Hidaka H, Endo T.  Selective inhibitors of three forms of cyclic nucleotide phosphodiesterases—basic and potential clinical applications. *Adv Cyclic Nucleotide Res* 16: 245-259, 1984.
5.  Lugnier C, Schoeffter P, LeBec A, Strouthou E, Stoclet JC.  Selective inhibition of cyclic nucleotide phosphodiesterases of human, bovine and rat aorta. *Biochem Pharmacol* 35:1743-1751, 1986.
6.  Ahn HS, Crim W, Romano M, Moroney S, Pitts B.  Effects of selective inhibitors on cyclic nucleotide phosphodiesterases (PDEs) of rabbit and pig aorta. *Pharmacologist* 29: 522, 1987.
7.  Hagiware M, Endo T, Hidaka H.  Effects of vinpocetine on cyclic nucleotide metabolism in vascular smooth muscle.  *Biochem Pharmacol* 33: 453-457, 1984.

8. Bergstrand H, Kristoffersson J, Lundquist B, Schurmann A. Effects of antiallergic agents, compounds 48/80, and some reference inhibitors on the activity of partially purified human lung tissue adenosine cyclic 3',5'-monophosphate and guanosine cyclic 3'5'-monophosphate phosphodiesterases. *Mol Pharmacol* 13: 38-43, 1977.

9. Ahn HS, Crim W, Romano M, Moroney SJ, Sybertz EJ, Pitts B. Calmodulin dependent phosphodiesterase (CaM-PDE) and cyclic GMP specific PDE (cG-PDE) distribution and response to inhibitors in aorta and non-vascular tissues. *Pharmacologist* 30: A213, 1988.

10. Hansen RS, Beavo AJ. Differential recognition of calmodulin-enzyme complexes by a conformation-specific anti-calmodulin monoclonal antibody. *J Biol Chem* 261: 14636-14645, 1986.

11. Ahn HS, Crim W, Romano M, Sybertz EJ, Pitts B. Effects of selective inhibitors on cyclic nucleotide phosphodiesterases of rabbit aorta. *Biochem Pharmacol* 38: 3331-3339, 1989.

12. Ahn HS, Eardley D, Watkins R, Prioli N. Effects of several newer cardiotonic drugs on cardiac cyclic AMP metabolism. *Biochem Pharmacol* 34: 1113-1121, 1986.

13. Ahn HS, Crim W, Pitts B, Sybertz EJ. Ca/CaM-stimulated and cGMP specific phosphodiesterases: tissue distribution, drug sensitivity and regulation of cGMP levels. *Adv Second Messenger Phosphoprotein Res*, Vol. 24, In Press, 1990.

14. Sharma RK, Wang JH. Isolation of bovine brain calmodulin-dependent cyclic nucleotide phosphodiesterase isozyme. *Methods in Enzymology* 159: 582-594, 1988.

# HEMODYNAMIC RESPONSE OF CONSCIOUS RATS AND DOGS

## TO THE PROTEIN KINASE C INHIBITOR STAUROSPORINE

R. Allan Buchholz, Ronald L. Dundore and Paul J. Silver

Department of Cardiovascular Pharmacology
Sterling Research Group
Rensselaer, NY 12144

## INTRODUCTION

Agonist-induced contraction of vascular smooth muscle appears to be mediated by $Ca^{2+}$ influx through voltage dependent and independent channels and by receptor-linked, G-protein coupled activation of phospholipase C (1). Two intracellular messengers, inositol trisphosphate ($IP_3$) and diacylglycerol (DAG), are generated by phospholipase C catalyzed hydrolysis of inositol bisphosphate (2). Considerable biochemical and physiological evidence suggests that $IP_3$-mediated mobilization of intracellular $Ca^{2+}$ results in calcium-regulated phosphorylation of the 20,000 dalton myosin light chain and the initiation of smooth muscle contraction (3-5). However, the mechanism(s) responsible for sustaining isometric force in vascular smooth muscle are not fully understood. Several mechanisms have been proposed for the maintenance of smooth muscle tension, including the "latch bridge" state (6) and DAG activation of protein kinase C (PKC) (7). Evidence supporting the role of PKC in the maintenance of smooth muscle tone is further reviewed in an earlier chapter (8). Staurosporine has been identified as a very potent PKC inhibitor (9). The cardiovascular actions of staurosporine *in vivo* have not been fully characterized. Therefore, we examined the hemodynamic response to staurosporine in conscious, spontaneously hypertensive rats (SHR) and conscious normotensive dogs.

## METHODS

### Spontaneously Hypertensive Rats

Male, SHR (Charles River) 15-17 weeks of age were anesthetized with sodium pentobarbital (50 mg/kg, i.p.) and heparin-filled (200 units/ml), teflon-tipped Tygon cannulas (10) were placed in the abdominal aorta and inferior vena cava via the left femoral artery and vein, respectively. After a two day recovery period, the rats were placed in individual Plexiglas testing boxes and the arterial cannula attached to a Statham pressure transducer (P23 ID) positioned at heart level. Pulsatile arterial pressure was recorded on a Grass polygraph and mean arterial pressure (MAP) was derived by electronic filtering of the pulsatile pressure signal. Heart rate was calculated from the systolic pressure peaks. Based upon preliminary experiments, staurosporine (or vehicle - PEG 400 and ethanol in saline 1:1:2) was administered intravenously in a single dose estimated to normalize the MAP of SHR (25-30% reduction). Cardiovascular activity was recorded continuously for 10 hours after medication to assess the duration of the depressor response.

**Normotensive Dogs**

Under aseptic conditions, 7 male and female mongrel dogs, 12-16 kg, were instrumented for chronic hemodynamic measurements. A Tygon cannula was positioned in the descending thoracic aorta for measurement of pulsatile and mean arterial pressure. A Konigsberg solid-state pressure transducer was placed in the left ventricle (LV) through a stab-wound in the apex of the heart for measurement of LV pressure and LV dP/dt, an index of myocardial contractility. A Transonic ultrasonic flow transducer was positioned on the ascending aorta for continuous measurement of aortic blood flow, an index of cardiac output. Total peripheral vascular resistance was calculated off-line by dividing mean arterial pressure by cardiac output. A second Transonic flow transducer was positioned on the left renal artery for determination of renal blood flow. Heart rate was determined from a lead II ECG. After a 2-week recovery period, conscious dogs were administered staurosporine or vehicle (PEG and ethanol in saline) in an ascending cumulative-dose fashion (3-130 µg/kg, i.v.). Each dog received staurosporine and vehicle in a randomized within subjects design. A one week clearance period was given between each test. The test agent was administered at increasing doses every 15 min, while cardiovascular activity was recorded continuously on a Grass polygraph.

RESULTS

The baseline MAP of the conscious SHR used in this study was 165 mmHg. As shown in Figure 1, a single 300 µg/kg intravenous injection of

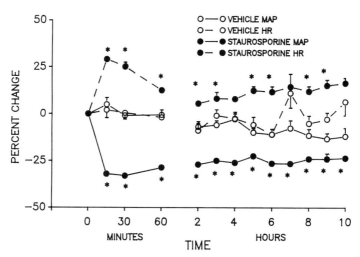

Figure 1: Mean arterial pressure (MAP) and heart rate (HR) responses of conscious SHR to a single intravenous injection of staurosporine (300 µg/kg) or vehicle (1 ml/kg). Values are means ± SEM expressed as percent change from baseline. n=8/group, * = $p < 0.05$ vs vehicle.

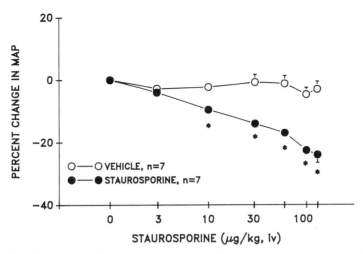

Figure 2: Mean arterial pressure response of conscious normotensive dogs to ascending cumulative doses of staurosporine or vehicle. Values are means ± SEM expressed as percent change from baseline. * = $p < 0.05$ vs vehicle.

staurosporine produced an initial 33% reduction in the MAP of SHR within 30 min. A persistent 25% reduction in the MAP of SHR given staurosporine was observed throughout the 10-hour recording period. The decrease in MAP caused by staurosporine was accompanied by a slight, but sustained, tachycardia. Oral administration of staurosporine to SHR also produced a gradual decrease in MAP, with a maximum reduction in MAP 6 hours after medication.

Staurosporine also caused a dose-dependent decrease in the MAP of conscious, normotensive dogs (Fig. 2). The decrease in MAP was associated with a significant 49% reduction in total peripheral resistance at the highest dose tested (130 µg/kg, Table 1). Significant dose-dependent increases in heart rate and cardiac output, most likely of baroreflex origin, compensated for the marked reduction in vascular resistance and limited the fall in MAP. Surprisingly, the reflex increase in heart rate due to the staurosporine-induced fall in MAP was not accompanied by an increase in myocardial contractility, as revealed by the lack of effect on LV dP/dt. However, a significant increase in renal blood flow was observed.

DISCUSSION

Recent reports have demonstrated that age-related increases in systolic blood pressure are paralleled by increases in PKC activity in the aorta and platelets of SHR (11,12). Moreover, the putative PKC inhibitor staurosporine antagonizes phorbol ester- and norepinephrine-induced vascular contractions and inhibits stretch-dependent vascular tone in isolated tissues (13). In the present study, we found that staurosporine produced a long-lasting antihypertensive effect when administered intravenously or orally to conscious SHR. These data are consistent with the hypothesis that heightened PKC activity may contribute to the elevation of vascular tone observed in hypertension.

The hemodynamic mechanisms mediating the antihypertensive action of staurosporine in SHR were not examined. However, staurosporine also caused dose-dependent decreases in the MAP of

## TABLE 1

### HEMODYNAMIC RESPONSE OF CONSCIOUS DOGS TO INTRAVENOUS ADMINISTRATION OF THE PROTEIN KINASE C INHIBITOR STAUROSPORINE

| Variable | Group | Baseline[a] | DOSE (μg/kg, iv) | | | | | |
|---|---|---|---|---|---|---|---|---|
| | | | 3 | 10 | 30 | 60 | 100 | 130 |
| | | | Percent Change from Baseline | | | | | |
| MPA (mmHg) | STAURO | 97 ± 3 | -4 ± 1 | -10 ± 2 | -14 ± 1* | -17 ± 2* | -23 ± 1* | -24 ± 3* |
| | VEHICLE | 98 ± 4 | -3 ± 2 | -2 ± 2 | -1 ± 3 | -1 ± 3 | -5 ± 2 | -3 ± 3 |
| HR (bpm) | STAURO | 98 ± 5 | 7 ± 4 | 7 ± 4 | 23 ± 5 | 48 ± 8* | 69 ± 9* | 83 ± 15 |
| | VEHICLE | 100 ± 13 | 6 ± 4 | 8 ± 6 | 14 ± 11 | 9 ± 8 | -2 ± 6 | -3 ± 4 |
| CO (L/min) | STAURO | 1.6 ± 0.2 | 4 ± 6 | 12 ± 11 | 30 ± 16 | 40 ± 16* | 60 ± 22* | 62 ± 21* |
| | VEHICLE | 2 ± 0.4 | 12 ± 9 | 1 ± 5 | 10 ± 8 | 7 ± 6 | -3 ± 6 | -4 ± 7 |
| TPR (mmHg/L/min) | STAURO | 67 ± 9 | -4 ± 5 | -11 ± 5 | -24 ± 7* | -33 ± 6* | -45 ± 6* | -49 ± 5* |
| | VEHICLE | 65 ± 16 | -5 ± 5 | 3 ± 5 | -5 ± 5 | -0.5 ± 6 | 9 ± 8 | 11 ± 7 |
| LVP (mmHg) | STAURO | 120 ± 5 | -3 ± 2 | -5 ± 1 | -6 ± 2 | -6 ± 2 | -8 ± 2 | -12 ± 3 |
| | VEHICLE | 128 ± 7 | -5 ± 1 | -2 ± 1 | -3 ± 2 | -1 ± 2 | -4 ± 2 | -6 ± 2 |
| LV dP/dt (mmHg/sec) | STAURO | 2466 ± 177 | 4 ± 5 | 3 ± 4 | 9 ± 5 | 12 ± 7 | 15 ± 7 | 9 ± 9 |
| | VEHICLE | 2757 ± 321 | 7 ± 4 | 4 ± 3 | 8 ± 4 | 6 ± 5 | 2 ± 4 | 2 ± 5 |
| RBF (ml/min) | STAURO | 79 ± 11 | 27 ± 14 | 37 ± 24 | 64 ± 39 | 51 ± 23* | 40 ± 17* | 58 ± 22* |
| | VEHICLE | 85 ± 10 | 1 ± 11 | -3 ± 6 | -5 ± 5 | -1 ± 5 | -4 ± 8 | -14 ± 11 |

Values are mean ± SEM, n = 7/group.

a - Baseline values are expressed in absolute units indicated for each variable.

* - p < 0.05 vs vehicle.

conscious, normotensive dogs. The hemodynamic profile of staurosporine in this model indicated that a significant reduction in total peripheral vascular resistance primarily accounted for the fall in MAP. This finding is consistent with the actions of an agent acting on the peripheral vasculature, as would be expected with a PKC inhibitor, rather than affecting cardiac output or circulatory volume. As with other agents that cause peripheral vasodilation, the staurosporine-induced fall in MAP was associated with a temporally-related tachycardia, most likely of reflex origin. An increase in cardiac contractility did not accompany the reflex tachycardia. Sympathetically-mediated increases in contractility, primarily of baroreflex origin, are associated with the administration of other vasodilators (hydralazine, $Ca^{2+}$ blockers) in the conscious dog. Thus, the absence of this effect after staurosporine suggests three possible explanations: 1) either that an increase in sympathetic tone to the heart is absent and the observed tachycardia is due to vagal withdrawal, 2) staurosporine selectively antagonizes the sympathetic inotropic response to the reduction in vascular tone, or 3) staurosporine directly suppresses myocardial contractility. Staurosporine also significantly increased renal blood flow by as much as 50%. This is in contrast to other vasodilators that either do not affect renal blood flow or may decrease it, e.g., calcium blockers, hydralazine, nitrovasodilators.

The findings from these studies suggest that a protein kinase C inhibitor such as staurosporine produces a hemodynamic profile that would be highly desirable in a novel antihypertensive agent, i.e., long duration of action, limited effect on cardiac contractility, and enhanced renal blood flow. Whether staurosporine achieves these actions exclusively through PKC inhibition is not clear since it also inhibits myosin light chain kinase at similar concentrations (8). Additional research with more selective protein kinase inhibitors is necessary to better elucidate the roles of myosin light chain kinase and PKC in the regulation of vascular tone and arterial blood pressure.

# REFERENCES

1.  Nishimura J, Khalil RA, van Breeman C. Agonist-induced vascular tone. *Hypertension* 13: 835, 1989.
2.  Berridge MJ. Inositol trisphosphate and diacylglycerol: Two interacting second messengers. *Ann Rev Biochem* 56: 159, 1987.
3.  Kamm KE, Stull JT. The function of myosin and myosin light chain kinase phosphorylation in smooth muscle. *Ann Rev Pharmacol Toxicol* 25: 593, 1985.
4.  Kamm KE, Stull JT. Regulation of smooth muscle contractile elements by second messengers. *Ann Rev Physiol* 51: 299. 1989.
5.  Moreland RS, Cilea J, Moreland S. Calcium and phosphorylation dependent regulation of vascular smooth muscle contraction. *In: Cellular and Molecular Mechanisms of Hypertension,* R.H. Cox (ed). New York: Plenum Publishing, (in press), 1990.
6.  Hai CM, Murphy RA. $Ca^{2+}$, crossbridge phosphorylation, and contraction. *Ann Rev Physiol* 51: 285, 1989.
7.  Rasmussen H, Takuwa Y, Park S. Protein kinase C in the regulation of smooth muscle contraction. *FASEB J* 1: 177, 1987.
8.  Silver PJ, Pagani ED, Cumiskey WR, Dundore RL, Harris AL, Lee KC, Ezrin AM, Buchholz RA. Calcium-regulated protein kinases and low $K_m$ cGMP phosphodiesterases: Targets for novel antihypertensive therapy. *In: Cellular and Molecular Mechanisms*

*of Hypertension*, R.H. Cox (ed). New York: Plenum Publishing, (in press), 1990.

9. Tamaoki T, Nomoto H, Takahashi I, Kato Y, Morimoto M, Tomita F. Staurosporine, a potent inhibitor of phospholipid/Ca++ dependent protein kinase. *Biochem Biophys Res Commun* 135: 397, 1986.

10. Buchholz RA, Nathan MA. Chronic lability of the arterial blood pressure produced by electrolytic lesions of the nucleus tractus solitarii in the rat. *Circ Res* 54: 227, 1984.

11. Murakawa K, Kohno M, Yasunari K, Yokokawa K, Horio T, Takeda T. Possible involvement of protein kinase C in the maintenance of hypertension in spontaneously hypertensive rats. *J Hypertension* 6(suppl 4): S157, 1988.

12. Takaori K, Itoh S, Kanayama Y, Takeda T. Protein kinase C activity in platelets from spontaneously hypertensive rats (SHR) and normotensive Wistar-Kyoto rats (WKY). *Biochem Biophys Res Commun* 141: 769, 1986.

13. Laher I, Bevan JA. Staurosporine, a protein kinase C inhibitor, attenuates Ca2+-dependent stretch-induced vascular tone. *Biochem Biophys Res Commun* 158: 58, 1989.

# EFFECTS OF BUFALIN ON RENAL VENOUS OUTFLOW, URINE

## FLOW AND NATRIURESIS IN THE ANESTHETIZED DOG

D. Eliades, M.B. Pamnani, B.T. Swindall, and F.J. Haddy

Department of Physiology
Uniformed Services University
Bethesda, MD 20814

## INTRODUCTION

Recent studies suggest that the endogenous digitalis-like substance (DLS) implicated in the pathogenesis of low renin hypertension in animals and humans may be a steroidal dienolide derivative. One of these derivatives, bufalin (aglycone), inhibits $Na^+,K^+$-ATPase (1) and has recently been shown in our laboratory to have some of the physiological characteristics expected of a DLS (2). For example, infusion of bufalin into the brachial artery of the anesthetized dog increases vascular resistance and blocks $K^+$ vasodilation while it potentiates norepinephrine vasoconstriction (Fig. 1). When bufalin is administered intravenously in the dog, increases in heart rate, dP/dt, and arterial blood pressure ensue (Fig. 2). In some animals (3/7) post-infusion natriuresis and diuresis occurs. Intravenous infusion in the rat likewise produces increases in blood pressure, heart rate, and dP/dt, and marked diuresis, natriuresis and kaliuresis (3) during and post-infusion. The present study in the anesthetized dog has been designed to examine the mechanism for the post-infusion natriuresis and diuresis seen with bufalin.

## METHODS

Two groups of six male dogs (average weight 22 kg) were used in these studies. All animals were fasted, pentobarbital-anesthetized, and intubated for mechanical ventilation. Blood gases were monitored using a Radiometer Model ABL3 (Radiometer Copenhagen, Copenhagen, Denmark) and were normalized through adjustments in respiratory rate and volume. The animals were placed in left lateral recumbency and the renal artery and vein exposed through a left flank incision. Heparin (10 mg/kg) was administered intravenously and a bypass shunt was interposed between the renal and femoral veins. The bypass was designed with a side port to allow renal venous blood to be diverted into a graduated cylinder for periodic measurement of renal venous outflow (RBF, ml/10 sec). A fine needle was inserted into the renal artery for infusion of bufalin or vehicle into the renal circulation, and the ureter was cannulated for measurement of urine flow and electrolyte excretion. After mean arterial blood pressure,

*Cellular and Molecular Mechanisms in Hypertension*
Edited by R.H. Cox, Plenum Press, New York, 1991

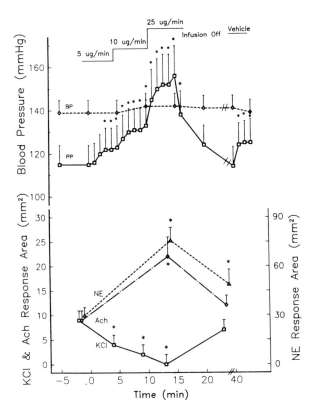

Figure 1: Effects of intra-arterial infusion of bufalin and vehicle (10% EtOH + 90% saline, 300 mOsm) on brachial artery perfusion pressure (PP) at constant blood flow (100 ml/min), mean aortic blood pressure (BP), and PP responses to IA injection of NE (0.1 μg), Ach (0.2 μg) and iso-osmotic KCl (1 ml) in the anesthetized dog. N = 7, *P ≤ 0.05 relative to control. From Eliades et al., 1989.

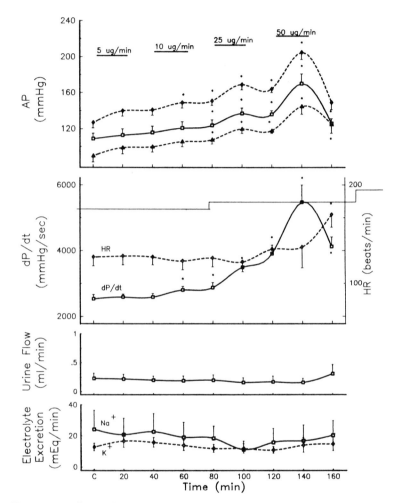

Figure 2: Effects of IV infusion of bufalin on systolic, diastolic and mean aortic pressures (AP), heart rate (HR), left ventricular dP/dt, urine flow and urinary Na+ and K+ excretion in the anesthetized dog. N=7, *P ≤ 0.05 relative to control. From Eliades et al., 1989.

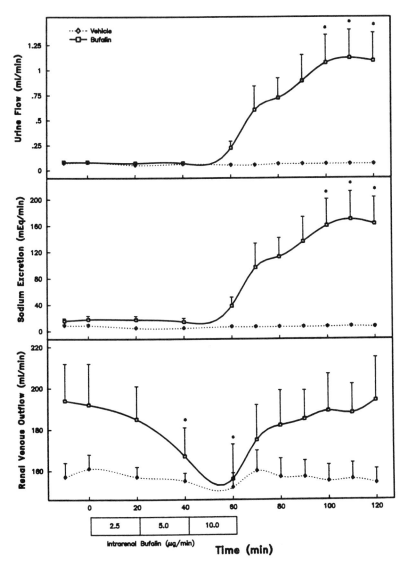

Figure 3: Effects of intrarenal infusions of bufalin and vehicle (10% EtOH + 90% saline, 300 mOsm) on urine flow, sodium excretion, and renal venous outflow in the anesthetized dog. N=6, *P ≤ 0.05 relative to control.

RBF and urine flow were stable for 20 min., bufalin (20 µg/ml) or vehicle (10% EtOH + 90% saline, 300 mOsm) was administered into the kidney in a stepwise manner at the three rates of 2.5, 5.0, and 10.0 µg/min (0.123, 0.247, 0.494 ml/min) for 20 min. each. Blood pressure, RBF and urine flow were measured every 10 min for an additional 60 min.

## RESULTS

Local intrarenal administration of bufalin resulted in a significant reduction in RBF during the second and third infusion rates relative to control. During this time, urine flow, sodium and potassium excretion, and mean arterial blood pressure remained constant (Fig. 3). The vehicle was without effect.

The renal vasoconstrictor effect of bufalin was reversed within 10 min after discontinuing the infusion. Forty min after ending the bufalin infusion there was a profound increase in sodium excretion and urine flow to 11 and 14 times control levels, respectively. This delayed natriuresis and diuresis was not seen in the vehicle-treated group.

## DISCUSSION AND CONCLUSIONS

Bufalin is a $Na^+,K^+$-ATPase inhibitor which belongs to a class of steroids with ouabain-like activity of animal rather than plant origin. It is almost identical structurally to resibufogenin, a compound recently identified in toad skin and plasma, and suggested to be a physiological regulator of $Na^+,K^+$-ATPase (4). The purified toad skin compound has been shown to bind to the ouabain receptor, inhibit $Na^+,K^+$-ATPase activity, and increase the force of contraction of cardiac tissue. Reports from many laboratories have suggested the presence of a DLS in plasma and tissues of various animals and man under the conditions of acute volume expansion and low renin hypertension (1,5).

This study extends previous work in our laboratory in both the dog and the Wistar rat (3) in which urine flow and sodium excretion increased most markedly in the post-infusion period. The post-infusion diuresis and natriuresis seen with bufalin infusion is the subject of the present work. In the present study we demonstrated that bufalin infusion into the canine kidney resulted in a significant decrease in renal venous outflow during the infusion, at which time urine flow and sodium excretion did not change. After the infusion was discontinued, urine flow and sodium excretion increased as renal venous outflow returned to control levels.

Bufalin has been shown to constrict the blood vessels in the forelimb of the dog (2), probably as a result of depolarization of vascular smooth muscle cells due to inhibition of the electrogenic $Na^+,K^+$ pump (7). The vasoconstriction is rapid in onset and offset, and is accompanied by decreased potassium-induced vasodilation and increased norepinephrine-induced vasoconstriction, as one may expect from a $Na^+,K^+$-ATPase inhibitor. We believe that it is possible that when bufalin is infused into the kidney, vasoconstriction occurs, decreasing renal perfusion and thereby negating any natriuresis due to tubular effects. Urine flow and sodium excretion would therefore be restricted during the bufalin infusion, and only increased after release of the vasoconstriction.

Thus, it appears that in the dog the diuretic and natriuretic capabilities of this Na⁺,K⁺-ATPase inhibitor are masked during infusion by renal vasoconstriction, and that these effects become apparent only after the bufalin infusion is stopped.

REFERENCES

1. Brownlee AA, Lee G, Maills IH. Marked inhibition of canine renal Na⁺,K⁺-ATPase by a bufodienolide but weak natriuretic activity in the rat. *J Physiol* 390: 94P, 1987.
2. Eliades D, Swindall B, Johnston J, Pamnani M, Haddy F. Hemodynamic effects of bufalin in the anesthetized dog. *Hypertension* 13: 690-695, 1989.
3. Pamnani MB, Eliades DC, Schooley JF, Haddy FJ. Effects of bufalin, a Na⁺,K⁺-ATPase inhibitor, on cardiovascular hemodynamics and renal excretory function in rats. *Circulation* 78(Suppl II): II-369, 1988.
4. Lichtstein D, Kachalsky S, Deutsch J. Identification of a ouabain-like compound in toad skin and plasma as a bufodienolide derivative. *Life Sci* 38: 1261-1270, 1986.
5. Haddy FJ, Overbeck HW. The role of humoral agents in volume expanded hypertension. *Life Sci* 19: 935-948, 1976.
6. Blaustein MP. Sodium ions, calcium ions, blood pressure regulation, and hypertension: A reassessment and a hypothesis. *Am J Physiol* 232: C165-C173, 1977.
7. Pamnani MB, Harder DR, Huot SJ, Bryant HJ, Kutyna FA, Haddy FJ. Vascular smooth muscle membrane potential and a ouabain-like humoral factor in one-kidney, one clip hypertension in rats. *Clin Sci* 63: 31s-33s, 1982.

# ROLE OF PHOSPHATIDYLINOSITOL TURNOVER

# IN THE CONTRACTION OF THE RAT AORTA

Evangeline D. Motley, Robert R. Ruffolo, Jr.
Douglas W.P. Hay and Andrew J. Nichols
Department of Pharmacology

SmithKline Beecham Pharmaceuticals
King of Prussia, PA 19406 and
Department of Physiology and Biophysics
Howard University College of Medicine
Washington, DC 20059

## INTRODUCTION

The contraction of vascular smooth muscle in the rat aorta has been shown to be mediated exclusively by $\alpha_1$-adrenoceptors located on the cell membrane. These $\alpha_1$-adrenoceptors mediate vasoconstriction by stimulating both the influx of extracellular calcium through membrane calcium channels and the release of intracellular calcium stores (1,2). Recently, it has been hypothesized that full $\alpha_1$-adrenoceptor agonists can activate $\alpha_1$-adrenoceptors and stimulate both the influx of extracellular calcium and the release of intracellular calcium, whereas, partial $\alpha_1$-adrenoceptor agonists may only cause the influx of extracellular calcium (1,3).

The stimulation of $\alpha_1$-adrenoceptors activates phospholipase C, which hydrolyzes phosphatidylinositol-4,5-bisphosphate to produce, among other things, inositol-1,4,5-trisphosphate and 1,2-diacylglycerol (4). Inositol-trisphosphate has been shown to cause the release of calcium from the sarcoplasmic reticulum in skinned muscle cells (5,6) and thus, may act as a mediator of the release of intracellular calcium.

In this study, the full $\alpha_1$-adrenoceptor agonist, (-)-norepinephrine, was used to look at the relationship betwen $\alpha_1$-adrenoceptor occupancy and the stimulation of phosphatidylinositol (PI) turnover and contractile response in the rat aorta.

## METHODS

### General

Normotensive male Sprague-Dawley rats (200-300 g) were sacrificed by a blow to the head. The thoracic aorta was removed and dissected free of fat and connective tissue in oxygenated physiological salt solution (PSS) of the following composition (in millimolar): NaCl, 119; KCl, 4.7; KH$_2$PO$_4$, 1.2; MgSO$_4$, 1.5; CaCl, 2.5; NaHCO$_3$, 25; glucose, 11; ascorbic acid 5; and disodium EDTA, 0.03; dissolved in demineralized water (pH 7.4 at 37°C). The PSS also contained cocaine (10 μM), propranolol (1 μM) and corticosterone (30 mM). The endothelium of the aorta was removed by rubbing the lumen with a blunt needle.

## Contraction

Aortic rings (3-5 mm long) were mounted in 10 ml organ baths containing PSS continuously aerated with a 95% O$_2$-5% CO$_2$ mixture maintained at 37°C. The tissues were attached to force displacement transducers for the measurement of isometric contractions and allowed to equilibrate at a resting tension of 1 g for 2 hr. Cumulative dose-response curves were constructed for (-)-norepinephrine (10$^{-10}$ - 10$^{-5}$ M) by increasing the bath concentration approximately 3-fold after the previous concentration had produced its maximal effect. All responses were expressed as a percentage of the response to (-)-norepinephrine (10$^{-6}$ M) obtained 1-2 hr before generation of the cumulative dose-response curve. Some tissues were incubated with the selective competitive α$_1$-adrenoceptor antagonist, prazosin (10$^{-8}$ M) for 30 min, or the irreversible α-adrenoceptor antagonist, phenoxybenzamine (10$^{-8}$ M), for 15 min followed by washing for 45 min, before construction of the dose-response curve.

## Measurement of [$^3$H]Phosphatidylinositol Hydrolysis

The vessels were placed in 2 ml PSS containing 10 μCi/ml [$^3$H]myo-inositol and incubated for 2 hr. Following incubation, the vessels were washed 3 times with 3 ml aliquots of PSS + 10 mM LiCl + 1 mM myo-inositol. Each vessel was cut into 8 rings and the rings were placed into glass tubes containing 990 μl of PSS + 10 mM LiCl. (-)-Norepinephrine (10$^{-9}$ - 10$^{-4}$ M) was added to the tubes and the tissues were incubated for 1 hr under 95% O$_2$-5% CO$_2$ and continuous shaking. Some tissues were incubated with prazosin (10$^{-7}$ M) for 30 min before incubation with (-)-norepinephrine, while others were incubated with phenoxybenzamine (3 x 10$^{-8}$ M) for 15 min and washed for 45 min before incubation with (-)-norepinephrine. The reaction was terminated with 3 ml of CHCL$_3$-MeOH (1:2) and subsequent addition of 1 ml of H$_2$O and 1 ml of CHCl$_3$. [$^3$H]Myo-inositol-monophosphate (IP), a major breakdown product of inositol-trisphosphate in the presence of LiCl, was isolated from the aqueous phase by ion-exchange chromatography using 1 ml of AG 1 x 8 resin. The columns were eluted sequentially with 16 ml of H$_2$O, 15 ml of 60 mM sodium formate - 5 mM sodium borate and 15 ml of 1 M ammonium formate - 0.1 M formic acid. [$^3$H]Myo-inositol-monophosphate eluted in the last wash was quantified by liquid scintillation spectrometry in DPM. The data was expressed as DPM/mg of dry tissue.

## Calculation of Dissociation Constants

The dissociation constants (K$_d$) for the contractile response and IP production were calculated by the Furchgott method (7) using the equation:

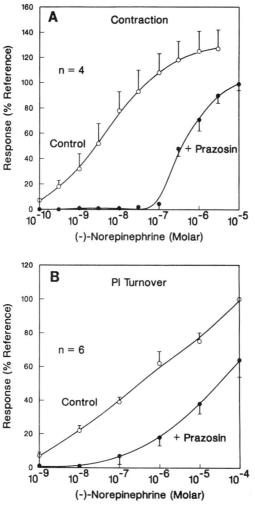

Figure 1: Dose-response curves to (-)-norepinephrine in the absence (○) and presence (●) of (A) prazosin ($10^{-8}$ M) for the contractile response and (B) prazosin ($10^{-7}$ M) for IP production in the rat aorta.

$$\frac{1}{[A]} = \frac{1}{[A']} \times \frac{1}{q} + \frac{1}{K_d} \times \frac{(1-q)}{q}$$

where [A] and [A'] are equipotent agonist concentrations before and after alkylation with phenoxybenzamine. $K_d$ was calculated by:

$$K_d = \frac{\text{slope - 1}}{\text{intercept}}$$

## Generation of Occupancy-Response Curves

The fraction of $\alpha_1$-adrenoceptors occupied (y) at a given concentration of (-)-norepinephrine [A] was calculated by the law of mass-action (8):

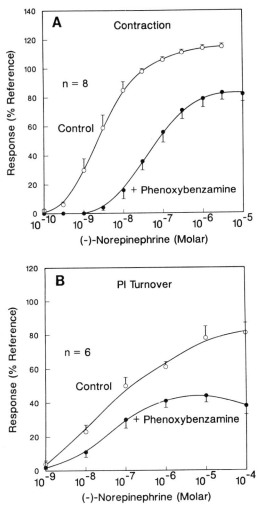

Figure 2: Dose-response curves to (-)-norepinephrine in the absence (O) and presence (●) of (A) phenoxybenzamine ($10^{-8}$ M) for the contractile response and (B) phenoxybenzamine ($3 \times 10^{-8}$ M) for IP production in the rat aorta.

$$y = \frac{[A]}{K_d + [A]}$$

For each (-)-norepinephrine concentration, the response was determined and plotted against the receptor occupancy.

## RESULTS

The full $\alpha_1$-adrenoceptor agonist, (-)-norepinephrine, produced a dose-dependent contractile response in the rat aorta which was sensitive to the $\alpha_1$-adrenoceptor antagonist, prazosin ($10^{-8}$ M), which produced a parallel rightward shift in the dose-response curve (Figure 1A). (-)-Norepinephrine also produced a dose-dependent increase in IP production which was competitively inhibited by prazosin ($10^{-7}$ M) (Figure 1B).

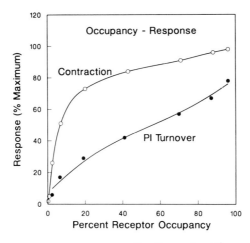

Figure 3: Occupancy-response curves for the contractile response (O) and
IP production (●) in the rat aorta.


The $EC_{50}$ of the contractile response to (-)-norepinephrine was 3 x $10^{-9}$
M. The irreversible $\alpha_1$-adrenoceptor antagonist, phenoxybenzamine ($10^{-8}$
M), produced a 13-fold reduction in the $EC_{50}$ with a 30% reduction in the
maximum response (Figure 2A). Using the Furchgott method, the
dissociation constant was calculated to be 3.8 x $10^{-8}$ M. Phenoxybenzamine
(3 x $10^{-8}$ M) produced a 46% reduction in the maximum IP production with
no change in $EC_{50}$ (Figure 2B). The dissociation constant calculated by the
Furchgott method was 4.3 x $10^{-8}$ M.

The law of mass-action was used to calculate the fraction of $\alpha_1$-
adrenoceptors occupied for the contractile response and the IP production
(Figure 3). The occupancy-response curve for the contractile response was
hyperbolic and showed that occupation of only 6% of the $\alpha_1$-adrenoceptor
population was required to produce the half-maximal response. The IP
production curve was nearly linear and showed that occupation of 45% of
the $\alpha_1$-adrenoceptors was required to produce the half-maximal stimulation
of IP production.


## DISCUSSION

This study was designed to investigate the relationship between $\alpha_1$-
adrenoceptor occupancy by (-)-norepinephrine and the stimulation of PI
turnover and contractile response in the rat aorta. The data show that (-)-
norepinephrine produced a prazosin sensitive, dose-dependent contractile
response and stimulation of IP production. Phenoxybenzamine produced
similar reductions in the maximum response of these two parameters but
had a differential effect on the $EC_{50}$, revealing different occupancy-response
relationships for these two effects. Thus, while there is a large $\alpha_1$-
adrenoceptor reserve for the contractile response to (-)-norepinephrine in
the rat aorta, there is very little or no $\alpha_1$-adrenoceptor reserve for the
production of IP. Since there is a linear relationship between IP production
and intracellular calcium release in rat aorta (9), it is likely that there is no
$\alpha_1$-adrenoceptor reserve for the release of intracellular calcium. Thus,
there must be a large $\alpha_1$-adrenoceptor reserve for the influx of extracellular

calcium in order to give a large overall receptor reserve for the contractile response. These findings support the hypothesis that a single $\alpha_1$-adrenoceptor is coupled to two signal transduction processes with different efficiencies of coupling and mediated via two distinct G-proteins (10). The activation of one G-protein leads to the influx of extracellular calcium with high efficiency, while activation of the other G-protein stimulates phospholipase C with low efficiency, to hydrolyze phosphatidylinositol-4,5-bisphosphate, which leads to the release of intracellular calcium.

# REFERENCES

1.  Chiu AT, McCall DE, Thoolen MJMC, Timmermans PBMWM. Calcium utilization in the constriction of rat aorta to full and partial $\alpha_1$-adrenoceptor agonists. *J Pharmacol Exp Ther* 238: 224-231, 1986.
2.  Ruffolo RR Jr, Nichols AJ. The relationship of receptor reserve and agonist efficacy to the sensitivity of $\alpha$-adrenoceptor-mediated vasopressor responses to inhibition by calcium channel antagonists. *Ann NY Acad Sci* 522: 361-376, 1988.
3.  Nichols AJ, Ruffolo RR Jr. The relationship of $\alpha$-adrenoceptor reserve and agonist intrinsic efficacy to calcium utilization in the vasculature. *Trends Pharmacol Sci* 9: 236-241, 1988.
4.  Timmermans PBMWM, Thoolen MJMC. Calcium utilization in signal transformation of $\alpha_1$-adrenergic receptors. In: *The $\alpha_1$-Adrenergic Receptors*, Ruffolo RR (ed), Clifton: Humana Press, pp 113-187, 1987.
5.  van Breemen C, Lukeman S, Leijten P, Yamamoto H, Loutzenhiser R. The role of superficial SR in modulating force development induced by calcium entry into arterial smooth muscle. *J Cardiovasc Pharmacol* 8(Suppl 8): S111-S116, 1986.
6.  Hashimoto T, Hirata M, Itoh T, Kanmura Y, Kuriyama H. Inositol-1,4,5-trisphosphate activates pharmacomechanical coupling in smooth muscle of the rabbit mesenteric artery. *J Physiol* 370: 605-618, 1986.
7.  Kenakin TP. The classification of drugs and drug receptors in isolated tissues. *Pharmacol Rev* 36: 165-222, 1984.
8.  Ruffolo RR Jr. Review: Important concepts of receptor theory. *J Auto Pharmacol* 2: 277-295, 1982.
9.  Chiu AT, Bozarth JM, Timmermans PBMWM. Relationship between phosphatidylinositol turnover and calcium mobilization induced by $\alpha_1$-adrenoceptor stimulation in the rat aorta. *J Pharmacol Exp Ther* 240: 123-127, 1987.
10. Nichols AJ, Motley ED, Ruffolo RR Jr. Effect of pertussis toxin treatment on postjunctional $\alpha_1$- and $\alpha_2$-adrenoceptor function in the cardiovascular system in the pithed rat. *J Pharmacol Exp Ther* 249: 203-209, 1989.

# REDUCED AORTIC AND ARTERIOLAR GROWTH BY CAPTOPRIL IN

# NORMOTENSIVE AND RENAL HYPERTENSIVE RATS

Duo H. Wang and Russell L. Prewitt

Department of Physiology
Eastern Virginia Medical School
Norfolk, Virginia 23501

## INTRODUCTION

The influence of angiotensin II (AII) on vascular structure has been widely studied. Much evidence from *in vitro* experiments indicate that AII stimulates DNA and protein synthesis in cardiovascular tissue and induces hypertrophy in rat and human vascular smooth muscle cells. *In vivo* experiments using converting enzyme inhibition in hypertensive rats have been confounded with pressure decreases which can also reduce vascular wall structure. This experiment was designed to evaluate the effect of the renin-angiotensin system on microvascular function and structure in normotension and renal hypertension, and to determine whether captopril alters cross-sectional area of the aortic and arteriolar wall. These are the first whole animal data to show effects on aortic and microvascular wall growth when the production of AII is blocked and the animals are still hypertensive.

## MATERIALS AND METHODS

Male Wistar rats were uninephrectomized and divided into four treatment groups at the age of 6-7 weeks: 1. CONTROL (n = 11); 2. CON-CAP (n = 11), control given captopril; 3. 1K1C-CAP (n = 11), renal artery stenosis and captopril; 4. 1K1C (n=5), renal artery stenosis. Captopril was given in the drinking water at a concentration of 380 mg/l which provides a dose of approximately 100 mg/kg/day. Using intravital microscopy in the cremaster muscle, 1st- (1A) through 4th-order (4A) arteriolar dimensions and the number of 4A's on a single 3A were measured 4 weeks later before and after topical application of $10^{-3}$ M adenosine. The cremaster arterioles were then filled with microfil and cleared with increasing concentrations of glycerin in water. Measured by stereological techniques, the arteriolar density was determined as length of vessels per unit area of muscle. Using histological techniques, the medial-intimal area and diameter of the abdominal aortae were measured by a video image analysis system.

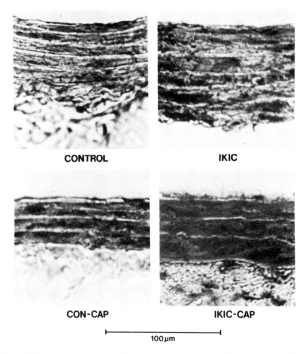

CONTROL                        IKIC

CON-CAP                     IKIC-CAP

100 μm

Figure 1: Light micrographs showing representative cross-sections of four abdominal aortae evaluated in these studies. The medial-intimal area was smaller in CON-CAP but larger in 1K1C compared to control. Captopril prevented hypertrophy of the wall in 1K1C-CAP.

RESULTS

Although captopril significantly decreased mean blood pressure in control rats from 124 ± 4 mmHg to 103 ± 5 mmHg, mean blood pressure was not significantly changed in treated (183 ± 5 mmHg) vs untreated hypertensive rats (193 ± 5 mmHg). Heart rates were similar in all groups.

Internal diameters (μm) of the abdominal aorta were 1282 ± 40, 1381 ± 66, 1273 ± 57 and 1356 ± 47 for CONTROL, CON-CAP, 1K1C-CAP and 1K1C, respectively. There were no statistical differences among the four groups. Medial-intimal area (mm²) of the abdominal aorta was significantly decreased in CON-CAP rats (0.22 ± 0.02) but increased in 1K1C rats (0.42 ± 0.01) compared to control (0.29 ± 0.01). Captopril reduced hypertrophy of the aortic wall significantly in 1K1C-CAP rats (0.31 ± 0.02). Representative micrographs of the vascular wall are shown in Figure 1.

There was a significant decrease in the relaxed diameters of 1A's and 2A's in hypertensive rats versus both control groups (Fig. 2). Captopril treatment significantly reduced the relaxed diameter of 1A's in CON-CAP versus control rats. There were no significant changes in diameters of 3A's and 4A's (Fig.2).

In spite of a significant increase in wall-to-lumen ratio of 1A's in CON-CAP (0.211 ± 0.008), 1K1C-CAP (0.318 ± 0.012) and 1K1C (0.388 ± 0.009) versus control (0.169 ± 0.004), there were significant decreases in cross-sectional wall area (CSWA) in 1A's, 2A's and 4A's in 1K1C-CAP and in 1A's in CON-CAP rats (Fig.2). As shown in previous experiments (1) and in Figure 2, CSWA was not reduced by hypertension alone.

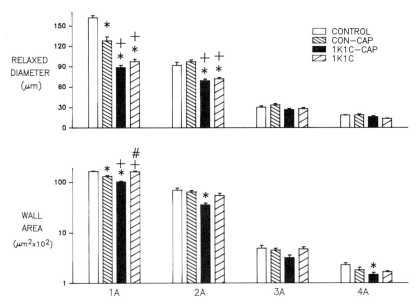

Figure 2: Upper panel: Internal diameter of 1st- through 4th-order (1A-4A) arterioles after dilation with topical $10^{-3}$ M adenosine. Lower panel: cross-sectional wall area of arterioles. Symbols indicate statistical significance at $p < 0.05$ versus control (*), CON-CAP (+) and 1K1C-CAP (#).

There were no significant differences in the number of flowing 4A's on a 3A under resting conditions. The number of open 4A's significantly increased after the application of adenosine in all groups. This number was significantly lower in the CON-CAP ($7.0 \pm 0.4$) and 1K1C-CAP ($6.8 \pm 0.5$) than in untreated controls ($9.2 \pm 0.6$), but unchanged in 1K1C ($8.4 \pm 0.2$).

Figure 3 shows the arteriolar density of large and small arterioles as total length per unit area of tissue. There were no differences in large arteriolar density. Arteriolar density for 3A's, 4A's and 5A's was significantly decreased in captopril treated groups and the 1K1C group. The important finding is that structural rarefaction of arterioles was not prevented by captopril. Rather, chronic treatment with large doses of captopril appeared to cause rarefaction of small arterioles in normotensive rats.

DISCUSSION AND CONCLUSIONS

Although 1K1C is not a renin-dependent model and the acute high renin phase lasts for no more than a week (2), captopril may inhibit local formation of AII produced by increased vascular wall renin activity in hypertension (3), as well as in normotension (4). As shown in Figure 2, the structural reduction of large arteriolar diameters was found not only in hypertensive rats but also in treated normotensive rats, which may be indicative of an inhibition of growth by captopril. Furthermore, the marked decreases in CSWA of aortae and arterioles in captopril treated rats are strong evidence that the growth of vascular smooth muscle cells was inhibited. Growth of the arterial wall is associated with elevated blood pressure (5). In the present study, however, prevention of hypertrophy of the aortic wall in 1K1C-CAP rats cannot be explained by altered mechanical

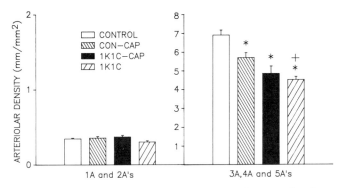

Figure 3: Density of 1st- through 5th-order (1A-5A) arterioles determined as length of vessels per unit area of muscle. Vessels were perfusion-fixed in the dilated state and filled with microfil. Symbols indicate statistical significance at p < 0.05 versus control (*) and p < 0.05 versus CON-CAP (+).

stress, because the blood pressure was not lowered by captopril. In addition, the CSWA's were significantly decreased in large and small arterioles of 1K1C-CAP and in 1A's of CON-CAP rats, suggesting the involvement of either AII or bradykinin in the regulation of vascular wall structure.

In recent years it has become clear that the AII is not only a potent vasoconstrictor but also a trophic or mitogenic factor which may play a direct role in the development of hypertension or chronic vascular disease. Blood vessels in a renin-secreting tumor with high local concentration of AII may be hypertrophic while vessels in nearby renal tissue, presumably exposed to a similar pressure but lower AII concentrations, are not (6). In Bartter's syndrome (7) and in familial chloride diarrhea (8) vascular hypertrophy develops in spite of normal or low blood pressure when renin and AII are markedly raised.

AII stimulates not only growth of the vascular wall, but also an increase in the number of vessels. Implantation of AII in the avascular rabbit cornea not only facilitated the activation of preexisting collateral vascular pathways, it also had angiogenic properties (9). This is consistent with the present finding that small arteriolar density was decreased in captopril treated rats.

The exact mechanism whereby AII influences vascular wall structure is not known. The acceleration of hydrolysis of polyphospho-inositide lipids by occupation of receptors by AII could increase DNA synthesis through activation of the $Na^+/H^+$ antiport (10). AII has been shown to induce the expression of *c-fos, c-myc*, and platelet-derived growth factor A-chain mRNA in cultured aortic smooth muscle cells (11). Alternatively, intracellular angiotensin may also bind to, and activate, potential nuclear receptors (12), leading to protein synthesis. Potentiation of sympathetic activity (13) by AII is another possible mechanism.

Another effect of captopril that cannot be ignored is the inhibition of bradykinin catabolism. Bradykinin, however, is a mitogenic agent (14), and increased concentrations would be expected to augment the proliferative response. However, there could be an as yet unknown mechanism whereby bradykinin may lead to decreased vascular growth.

In conclusion, captopril treatment resulted in an inhibition of vascular growth, both in size and number. This effect is possibly due to the blockade of AII production because AII is known to have hypertrophic and angiogenic effects.

## ACKNOWLEDGEMENTS

Supported in part by NIH Grant HL-36551.

## REFERENCES

1. Hashimoto H, Prewitt RL, and Efaw CW. Alterations in the microvasculature of one-kidney, one-clip hypertensive rats. *Am J Physiol* 253: H933-H940, 1987.
2. Koletsky S, Pavlicko KM, Rivera-velez JM. Renin-angiotensin activity in hypertensive rats with a single ischemic kidney. *Lab Invest* 24: 41-44, 1971.
3. Asaad M, Antonaccio MJ. Vascular wall renin in spontaneously hypertensive rats. *Hypertension* 4: 487-493, 1982.
4. Mizuno K, Nakamaru M, Higashimori K, Inagami T. Local generation and release of angiotensin II in peripheral vascular tissue. *Hypertension* 11: 223-229, 1988.
5. Owens GK, Schwartz SM. Alterations in vascular smooth muscle mass in the spontaneously hypertensive rat. *Circ Res* 51: 280-289, 1982.
6. Lindop GM, Lever AF. Anatomy of the renin-angiotensin system in the normal and pathological kidney. *Histopathology* 10: 335-362, 1986.
7. Fujita T, Sakaguchi H, Sibagaki M, Fukui T, Nomura M, Sekiguichi S. Functional and histological studies. *Am J Med* 63: 467-474, 1977.
8. Pasternack A, Perheentupa J. Hypertensive angiopathy in familial chloride diarrhoea. *Lancet* 2: 1047-1049, 1966.
9. Fernandez LA, Twickler J, Mead A. Neovascularization produced by angiotensin II. *J Lab Clin Med* 105: 141-145, 1985.
10. Heagerty AM, Ollerenshaw JD. The phosphoinositide signalling system and hypertension. *J Hypertension* 5:515-524, 1987.
11. Naftilan AJ, Pratt RE, Dzau VJ. Angiotensin II induction of c-fos, c-myc and platelet derived growth factor in vascular smooth muscle cells. *Clin Res* 36:303A, 1988.
12. Re RN, Vizard DL, Brown J, Bryan SE. Angiotensin II receptors in chromatin fragments generated by micrococcal nuclease. *Biochem Biophys Res Commun* 119:220-227, 1984.
13. Antonaccio MJ, Kerwin L. Pre- and postjunctional inhibition of vascular sympathetic function by captopril in SHR. *Hypertension* 3(supp I): I-54-I-62, 1981.
14. Owen NE, Villereal ML. Lys-bradykinin stimulates Na$^+$ influx and DNA synthesis in cultured human fibroblasts. *Cell* 32:979-985, 1983.

# STRUCTURE AND FUNCTION OF THE ADRENERGIC

## RECEPTOR FAMILY

Neil S. Roth, Robert J. Lefkowitz and Marc G. Caron

From the Howard Hughes Medical Institute
Departments of Cell Biology, Medicine, and Biochemistry
Duke University Medical Center
Durham, North Carolina 27710

## INTRODUCTION

The interaction of hormones and drugs with their respective targets has been widely studied with the hope that a better understanding of the molecular basis of their actions would provide insights not only into the nature of their specific mechanisms but those involved in certain pathophysiologic states. Receptors represent the central locus of interaction between ligands/drugs and cells. Adrenergic receptors because of their ubiquitous and well defined effector mechanisms have been excellent models for the study of these processes. Catecholamines and various synthetic analogs bind to adrenergic receptors that are integral membrane proteins and lead to the generation of intracellular second messengers culminating in a physiologic response. As with many other types of receptors this cascade of events is mediated by specific effector molecules which are coupled to adrenergic receptors in the plasma membrane. These intermediary signal transducing proteins are called guanine nucleotide regulatory proteins or G-proteins because they bind and hydrolyze guanine nucleotide triphosphate (1).

There are several distinct types of adrenergic receptors whose classification and subclassification has classically been based upon pharmacological specificity and physiological action. These include the $\alpha_1$, $\alpha_2$, $\beta_1$ and $\beta_2$ adrenergic receptor subtypes (2). Recent developments in the molecular characterization of these receptors has lead to the identification of several additional subtypes including $\beta_1$, $\beta_2$, $\beta_3$; $\alpha_{1A}$, $\alpha_{1B}$, $\alpha_{1C}$ and $\alpha_{2A}$, $\alpha_{2B}$ and $\alpha_{2C}$ (3-7). These distinct receptor subtypes differ not only in ligand binding specificity, but they are coupled to different G-proteins and in turn have different second messenger systems. $\beta$-adrenergic receptors are coupled to the G-protein, $G_s$, and agonist activation of receptor leads to the formation of its second messenger, cyclic adenosine monophosphate (cAMP), via stimulation of the enzyme adenylyl cyclase. $\alpha_2$-Adrenergic receptors inhibit adenylyl cyclase through the inhibitory regulatory G-protein, $G_i$. The activation of the $\alpha_1$-adrenergic receptors leads to the generation of two intracellular second messengers, diacylglycerol and inositol triphosphate, as a result of the stimulation of the enzyme phospholipase C (2). The G-protein involved in this cascade has not been well characterized.

The cascade of events that begins with agonist occupancy of G-protein coupled receptors and leads to the generation of second messengers elicit various cellular responses, including, for example, the activation of specific kinases and subsequent phosphorylation of proteins (8), release of $Ca^{++}$ from intracellular stores (9), activation of ion channels (10,11) etc. There is a wealth of physiologic responses that are linked to the adrenergic receptors and this is because of the wide range of receptor subtypes and effector systems. Additionally, recent data suggests that individual receptor subtypes may have the capacity to be linked to different G-proteins (12-14) illustrating the diversity and complexity of signal transduction.

G-proteins are coupled to many different types of receptors, for hormones, drugs and neurotransmitters. Probably the best characterized of these systems is the light transduction system, in which rhodopsin, the receptor, upon activation by light interacts with the G-protein transducin to activate the enzyme c-GMP phosphodiesterase producing increased level of GMP and light perception (15). Perhaps there are other sensory receptors, e.g. odorent or taste (16) that may also be part of this family of proteins. Of all these different receptors for hormones and neurotransmitters, more is known about the family of adrenergic receptors than any other. All of the previously pharmacologically characterized subtypes ($\alpha_1$, $\alpha_2$, $\beta_1$, $\beta_2$) have been purified to homogeneity and reconstituted with G-proteins in phospholipid vesicles (17). Through the techniques of covalent modification, site-directed mutagenesis and chimeric receptor construction, the function and regulation of the adrenergic receptors has been extensively studied. Genes and/or complementary DNAs (cDNAs) for at least nine subtypes have been isolated, sequenced and expressed in mammalian cell lines (Table 1). In this review, we will discuss how the structure of the adrenergic receptors relates to their various functions such as ligand binding, G protein activation and receptor mediated desensitization.

## TOPOGRAPHICAL ORGANIZATION OF ADRENERGIC RECEPTORS

Understanding of the structure of G-protein coupled receptors is essential to understanding their function. Essentially, G-protein coupled receptors of which the adrenergic receptors are a prototype mediate two main functions: they bind ligands from the outside of the cell and they activate an effector system leading to the generation of intracellular second messengers capable of activating various biochemical processes. The primary structure of adrenergic receptors suggests that these are indeed integral membrane proteins. Table 1 summarizes the properties of the various subtypes of $\beta$, $\alpha_1$ and $\alpha_2$-adrenergic receptors that have obtained from the elucidation of the primary sequence of these receptors (2,3,6,7). These receptors all consist of a single subunit which comprises several

## Table 1
### Molecular properties of the family of adrenergic receptor subtypes

ADRENERGIC RECEPTORS: STRUCTURE AND GENES

| Mammalian Subtype | $M_r$ kDa | Peptide Length | Topology Identity (MSD) (%) | Introns | Chromosomal Location |
|---|---|---|---|---|---|
| $\beta_2$ | 64 | 413 | | 0 | 5 q31–q32 |
| $\beta_1$ | 64 | 477 | 71($\beta_2$) | | 10 q24–q26 |
| $\beta_3$ | – | 402 | 63($\beta_2$) | 0 | – |
| $\alpha_2$(C–10) | 64 | 450 | 42($\beta_2$) | 0 | 10 q24–q26 |
| $\alpha_2$(C–4) | 75 | 461 | 39($\beta_2$), 75($\alpha_{2C}$–10) | – | 4 – |
| $\alpha_2$(C–2) | – | 450 | 36($\beta_2$), 74($\alpha_{2C}$–10) | 0 | 2 – |
| $\alpha_{1B}$ | 80 | 515 | 42($\beta_2$) | 1 | 5 q31–q32 |
| $\alpha_{1C}$ | – | 466 | 42($\beta_2$), 72($\alpha_{1B}$) | – | 8 – |
| $\alpha_{1A}$ | – | 560 | 43($\beta_2$), 88($\alpha_{1B}$) | – | – – |

stretches (7) of 20-28 hydrophobic amino acids, which potentially represent membrane-spanning $\alpha$-helixes. Within these transmembrane domains the adrenergic receptors contain considerable amino acid sequence identity. The highest identity (~70%) is usually found among members of the same subfamily (i.e. $\beta_1$, $\beta_2$ and $\beta_3$; $\alpha_{1A}$, $\alpha_{1B}$, $\alpha_{1C}$). Within members of the adrenergic receptor family (i.e. $\beta$ vs $\alpha_1$ or $\beta$ vs $\alpha_2$), the identity within the transmembrane segments fall to about 45% slightly higher that the 30-40% identity revealed when comparing transmembrane amino acid sequences across large families such as the muscarinic, serotonergic and dopaminergic receptors. The regions of greatest diversity even among related receptor subtypes are the extracellular amino terminus; the third cytoplasmic loop as well as the carboxyl terminal. Both of these are particularly variable in length and composition. Within a family such as the $\beta$-adrenergic receptors the first and second intracellular loops are usually fairly well conserved. Other similarities shared among the G-protein coupled receptors involved one or more sites of extracellular N-linked glycosylation near the amino terminus, as well as several potential sites of regulatory phosphorylation in the cytoplasmic domains.

The exact membrane configuration of receptors of this group is speculative. The confirmation of the topography of this family of receptors remains difficult and has been based primarily upon hydropathicity plots and through comparison with bacteriorhodopsin. Henderson and Unwin (18) have used high resolution electron diffraction to establish the presence of a bundle of seven membrane spanning $\alpha$-helices in bacteriorhodopsin of Halobacterium halobium. Similar structural features have been established for vertebrate rhodopsin using similar techniques. A schematic model representing the sequence of the $\beta_2$-adrenergic receptor within the plasma membrane is presented in Fig. 1.

Unfortunately, the limited quantities of proteins for any other G-protein coupled receptor has been the major obstacle in confirming this

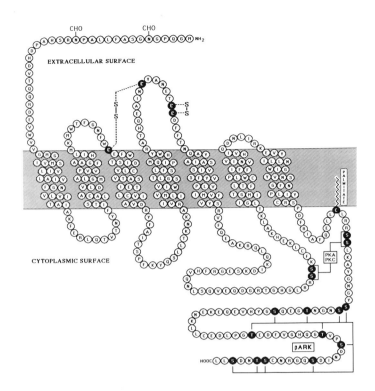

Fig. 1    Primary amino acid sequence of the human β₂-adrenergic receptor
and schematic representation of the organization of G-protein coupled
receptors within the plasma membranes.    Also highlighted are cysteine
residues putatively involved in disulfide bonds.    Cys341 is indicated in its
palmitoylated form.    Blackened serine and threonine residues represent
putative sites of phosphorylation by various kinases.

structural motif.    However through limited proteolysis Dohlman, et al. (19)
have assessed various features of this model and have confirmed many
elements.    Bahouth et al. (20)    have shown that antibodies to various
extracellular and intracellular loops of the β₂-adrenergic receptor were able
to interact with receptor in a fashion that was consistent with the seven
transmembrane domain topography.    Ultimate structural analysis will
undoubtedly require physical methods or X-ray crystallography of these
proteins.

## STRUCTURAL DOMAINS OF G-PROTEIN COUPLED RECEPTORS

The genes and/or cDNAs for a number of receptors coupled to G-
proteins have been sequenced and their amino acids deduced.    One of the
major goals of receptor research now is to determine which structural
domains of these receptors are responsible for the known functions of
receptors.    We will review the functions of the extracellular domains, the
transmembrane domains, the cytoplasmic domains and their roles in such
functions as ligand binding, G-protein interactions and functional
regulation of adrenergic receptors.

226

## Extracellular Domains

The extracellular domains of the G protein coupled receptors contain three loops that connect the seven hydrophobic transmembrane α-helixes as well as glycosylated amino terminus. The functions of the asparagine linked glycosyl residues of the amino termini is not known, however, several studies have demonstrated that they are not essential for ligand binding (21,22). The extracellular loops contain several cysteine residues which through the formation of disulfide bonds, may stabilize the ligand binding domain of these receptors (21,23,24). These residues are conserved in several G protein coupled receptors. Site directed mutagenesis of these cysteine residues conducted by Dixon et al. (21) showed that ligand binding was significantly altered when cysteine was substituted for valine at Cys106 and Cys184, two highly conserved extracellular cysteines. Dohlman et al. (24) found that substitutions of any of the four extracellular cysteines of β-adrenergic receptor (Cys 106, 184, 190, 191) lowered ligand binding and altered binding specificity. Interestingly, substitution of the four cysteine residues (Cys 77,116,125,285) within the hydrophobic transmembrane domains of the receptor, where ligand binding is thought to occur did not affect ligand binding (see below). Hence, it appears that the four extracellular cysteine residues (Cys106, 184, 190,191) of the $\beta_2$-adrenergic receptor ($\beta_2$-AR) may form disulfide bonds and contribute to stabilizing the ligand binding pocket. Rhodospin appears to share homology with the $\beta_2$-AR in this respect (25).

## The Transmembrane Domains

There is strong evidence that implicates the conserved transmembrane regions of the adrenergic receptors, in the binding of adrenergic ligands. This hypothesis has been supported by three different series of experiments using either photoaffinity labeling, site directed mutagenesis as well as receptor chimera. In several different types of G-protein coupled receptors, radioactive photoaffinity probes have been utilized and upon complete enzymatic digestion found to covalently label peptides predicted to reside within transmembrane domains. Dohlman et al. (26) showed that a point of covalent attachment of such a ligand in the $\beta_2$-adrenergic receptor was within the second membrane spanning region. Matsui et al. (27) localized the covalent attachment of a photoaffinity ligand in the $\alpha_2$-AR to the fourth transmembrane region. Additionally, photoaffinity labeling has been utilized in the muscarinic cholinergic (28), the $\beta_1$ receptor (29), and rhodopsin (30,31) receptor systems to localize the covalent attachment site of specific ligands to the transmembrane domains.

Various truncations and deletion mutations using site directed mutagenesis have further implicated the transmembrane region in the ligand binding function of receptors. Dixon et al. (21,32) have shown that hydrophilic cytoplasmic or extracellular loop deletions in the hamster $\beta_2$-AR, did not induce any changes in ligand binding while deletions within the hydrophobic domain altered ligand binding. Additionally, Kobilka et al. (33) showed that various β receptor mutants truncated in the seventh or after the fifth membrane spanning regions did not bind ligands. Additionally, substitutions of highly conserved amino acid residues in the first, second, third or seventh transmembrane domain could dramatically reduce ligand binding (34-36).

Receptor chimera further supports the fact that the transmembrane areas are involved in ligand binding. Chimera created using $\beta_1$-$\beta_2$-adrenergic receptor (37) showed that agonist specificity was mostly dependent on the fourth transmembrane domain. Chimeric $\alpha_2$-$\beta_2$-AR suggested that the sixth and seventh transmembrane regions were the most crucial in antagonist ligand binding (38). Strader et al. (39) have shown that two serine residues in the 5th transmembrane domain of the $\beta_2$AR are likely involved in hydrogen bonding the catechol group of catecholamines. It is interesting that these two serine residues are absolutely conserved in every catecholamine receptor so far cloned.

**TheCytoplasmic Domains**

The cytoplasmic domains of the adrenergic and related receptors consist of three intracellular loops and the carboxyl terminus. These regions are implicated in several functions, including G protein coupling as well as various covalent modifications. Both biochemical and mutagenesis techniques have been utilized to gain information into the function performed by these various domains. The most conserved cytoplasmic regions among G protein receptors are the first two cytoplasmic loops. Hence most studies to date have been directed towards the characterization of the regions with the most divergence, the third cytoplasmic loop and the carboxyl terminus.

Several studies have suggested that regions within the third cytoplasmic loop of the $\beta_2$-adrenergic receptor are integral part of the determinants of G proteins coupling. Strader and colleagues (40), as well as O'Dowd et al. (41) have used site directed and deletion mutagenesis to suggest that two regions within this third cytoplasmic loop, the amino and carboxyl ends adjacent to transmembrane 5 and 6, respectively, are implicated in coupling $\beta$ receptors to $G_s$. Results consistent with this view were also obtained using the chimeric receptor approach (38).

The N-terminal region of the carboxyl tail in both the $\beta_2$-adrenergic and rhodopsin appears to also play a role in receptor G protein coupling. O'Dowd et al. (42) have shown that a cysteine residue Cys341 of the human $\beta_2$-adrenergic receptor, which is highly conserved in most G protein coupled receptors, is palmitoylated and upon substitution of this residue, a nonpalmitoylated form of the receptor is created. This mutated $\beta_2$-adrenergic receptor shows a demonstrable decrease in receptor mediated adenylyl cyclase stimulation. Analogously, Cys322 and Cys323 of rhodopsin have been suggested to undergo fatty acylation with palmitic acid (44). It has been suggested that palmitoylation of these receptors and perhaps other G protein coupled receptors may serve to anchor this region of the receptor to the plasma membrane allowing for a conformation conducive to G protein coupling. Thus, data from site directed and substitution mutagenesis as well as chimeric receptor studies support the hypothesis that regions of the third cytoplasm loop and the cytoplasmic tail are involved in a receptor/G protein coupling. These cytoplasmic domains may also participate in a number of other receptor function such as receptor sequestration and down-regulation, however, further studies will need to be performed to establish this. These regions of receptor also appear to be involved in the processes of desensitization via covalent modification. This issue is discussed below.

DESENSITIZATION OF RECEPTOR FUNCTION

The functions and localization of receptors and other components of signal transduction systems are subject to dynamic regulation by a variety of mechanisms. The mechanisms by which these changes arise might occur in response to other hormonal signals such as glucocorticoids or thyroid hormones (44). However, and even much more direct way of controlling responsiveness involves the regulatory processes that occur upon prolonged exposure to an agonist. This phenomenon has been referred to as desensitization or tachyphylaxis. Desensitization is defined as the general process in which cellular response to a stimulus attenuates with time despite the continued presence of the stimulus. The mechanism of desensitization has been most thoroughly characterized in the β–adrenergic receptor system.

Two major patterns of desensitization have been historically defined. These patterns termed homologous desensitization and heterologous desensitization, differ as to the need for agonist occupancy of the receptor. In homologous desensitization of the β-adrenergic receptor exposure to a β-agonist leads to a diminished responsiveness only to subsequent stimulation by β-agonists, while stimulation of the adenylyl cyclase cascade by distinct types of agonists remains intact. On the other hand, heterologous desensitizaton is a more general blunting of responsiveness to multiple agonists working via distinct receptors that occurs after exposure of a cell to a particular agonist. However, now that we understand in some details some of the biochemical mechanisms implicated, these definitions are less useful. In the case of the β-adrenergic receptor as well as other receptors, both heterologous and homologous desensitization occur simultaneously and through a variety of mechanisms.

For G-protein coupled receptor blunting of responsiveness could involve regulatory modification of the receptor, the G-protein or the effector itself such as the adenylyl cyclase. Although evidence exist, still very little is understood about regulation of function of G-proteins and the adenylyl cyclase enzyme (45). Much more work has been done with respect to the mechanisms regulating receptor function. At the level of the receptor alone, three distinct types of regulatory mechanisms have been characterized. They can be distinguished on a temporal basis. These three mechanisms involved a rapid (seconds to minutes) uncoupling of the receptor response, a sequestration of the receptor (over several minutes) away from the accessibility of the cell membrane, and a longer (minutes to hours) down-regulation of the receptor translated by an actual loss of receptor molecules from the plasma membrane.

The rapid uncoupling of receptor function appears to be mediated by phosphorylation of the receptor by at least two distinct kinases, cAMP dependent protein kinase (PKA) and cAMP independent receptor kinase termed β-adrenergic receptor kinase (45,46). Initial evidence for the phosphorylation of the β-adrenergic receptor indicated that PKA actually phosphorylated the receptor. Phosphorylation of receptor correlated with the time course and dose response of agonists to induce desensitization and could be mimicked by cAMP analogs (47). Purified receptor was actually shown to be a substrate for PKA and that modification impaired its functionality (48,49). The suggestion that another kinase might be involved in this process came from the observation that agonist mediated desensitization of the β-receptor response in the mutant S49 lymphoma cells kin- and cyc- which respectively lack PKA and the $G_{s\alpha}$ subunit correlated with receptor phosphorylation (50). An enzyme activity, originally characterized in kin-S49 cells capable of phosphorylating the β-adrenergic receptor upon agonist occupancy was eventually isolated and purified from bovine brain (51). The enzyme, coined β-adrenergic receptor kinase (βARK), was capable of multiply phosphorylating (~8 moles/moles receptor) agonist occupied receptor reconstituted in lipid vesicles. A cDNA for βARK has been isolated from bovine brain (52). The enzyme is a protein of 689 amino acids with a centrally located kinase catalytic domain which bears similarities to other signal transduction kinases such as protein kinase A and C.

Although, phosphorylation of the β-adrenergic receptor by βARK decreases the receptor function, Benovic et al. (53) have suggested that other proteins or cofactors may be involved to fully achieve this reduction in activity. Comparing the effects of βARK, in both pure and crude form, one sees a dramatic difference in the ability of phosphorylated β-adrenergic receptors to interact with $G_s$. With purified βARK only a slight reduction in agonist stimulated GTPase activity of $G_s$ in a reconstituted system is noted while a crude βARK preparation reduces the above activity by 40-50% of its original activity (53). These findings suggested the potential presence of a factor which enhances the uncoupling of phosphorylated β-adrenergic receptor from its interaction with $G_s$. Interestingly, in the rhodopsin signal transduction system, such a cofactor, termed arrestin, is implicated in enhancing the effect of rhodopsin phosphorylation by rhodopsin kinase (54). Upon addition of arrestin to the reconstituted systems of the β-adrenergic receptor with pure βARK, the uncoupling of phosphorylated receptor from $G_s$ is restored to approximately 40-50% of its original level (53).

Following the initial isolation of a cDNA for the light transduction component arrestin (55), Lohse et al. (56) have isolated from bovine brain, a cDNA encoding a protein which shares significant homology (~60%) with arrestin. This protein named β-arrestin when expressed and partially purified is much more active at potentiating the uncoupling effect of phosphorylation of the $\beta_2$-adrenergic receptor than arrestin. In addition, β-arrestin is much more potent on the β-receptor system than the phosphorylated rhodopsin system and vice versa (56). Thus, β-arrestin appears to be a component of the hormone signalling system which acts as the arrestin component of the light transduction system to amplify the uncoupling effect of receptor phosphorylation (46).

The effects of βARK on the β-adrenergic receptor are analogous to the effects that rhodopsin kinase exerts upon rhodopsin. Rhodopsin kinase is an enzyme localized to the rod outer segments and it phosphorylates rhodopsin on a serine and threonine rich carboxyl terminus in a light dependent fashion (54). Phosphorylated rhodopsin and arrestin interact and decrease the ability of rhodopsin to activate transducin thereby essentially turning off the light transduction process. Interestingly, while each enzyme prefers its own substrate there is considerable ability of $\beta$ARK and rhodopsin kinase to interact with each other respective substrate. $\beta$ARK is capable of phosphorylating light bleached rhodopsin and conversely, rhodopsin kinase although weakly, is capable of phosphorylating agonist-occupied $\beta$-adrenergic receptors (57). $\beta$ARK is additionally capable of phosphorylating the agonist occupied form of the $\alpha_2$-adernergic receptor, in magnitude equal to that of the $\beta_2$-adrenergic receptor, but is not able to phosphorylate the $\alpha_1$-adrenergic receptor which unlike the $\beta_1$, $\beta_2$ and $\alpha_2$-adrenergic receptors is not coupled to adenylyl cyclase but to phosphatidylinositol phosphate hydrolysis. Hence, $\beta$ARK may be a general adenylyl cyclase coupled receptor kinase. The effects of $\beta$ARK may be mediated through agonist induced conformational changes that expose previously unavailable $\beta$ARK consensus sites on the cytoplasmic domains of the receptor. Clearly, the availability of molecular probes for $\beta$ARK will allow these questions to be addressed (52).

Recently, several experiments using site directed mutagenesis of the $\beta_2$-adrenergic receptor gene and the use of inhibitors of protein kinase A and $\beta$ARK in permeabilized whole cells have served to delineate the region of the receptor involved in regulating receptor function by phosphorylation and the relevance of these phosphorylation processes to the phenomenon of desensitization. There are two consensus sites for putative phosphorylation by PKA in the human and hamster $\beta_2$-adrenergic receptors. One is located in the carboxyl terminal portion of the 3rd cytoplasmic loop whereas the other is on the portion of the carboxyl terminal tail proximal to the seventh transmembrane domain. By analogy with rhodopsin, there are 11 serine and threonine residues that are concentrated toward the carboxyl terminal tail of the receptor and that are presumed to be the sites of phosphorylation by $\beta$ARK.

Hausdorff et al. (58) studied desensitization in Chinese hamster fibroblast cells using three mutant $\beta_2$-adrenergic receptor genes encoding receptors lacking putative phosphorylation sites for the cAMP dependent protein kinase A and/or the cAMP-independent $\beta_2$-adrenergic receptor kinase. It was shown that exposure of these cells to low concentrations (nanomolar) of agonist preferentially induce phosphorylation at protein kinase A sites as measured by a decreased sensitivity to agonist stimulation of adenylyl cyclase. Meanwhile, at higher concentrations (micromolar) phosphorylation on both the $\beta$ARK and protein kinase A sites occurs with maximal responsiveness of adenylyl cyclase markedly reduced. Hausdorff et al. (8) concluded that low or high concentrations of agonist elicit phosphorylation of $\beta$-adrenergic receptors on different domains, with different implications for the functional coupling of the receptor with effector molecules. For instance, physiologically these two biochemical mechanisms of desensitization may function in different situations one operating at low concentrations of agonist and another to high concentrations. One may speculate that protein kinase A which appears more active at low concentrations of agonist, may be more active in peripheral tissues, which are predominantly exposed to low concentrations of circulating catecholamines. Additionally, a desensitizing mechanism involving βARK may be more active at high concentrations of catecholamines, such as at synaptic locations. Interestingly, the brain is an excellent tissue source for βARK (51). Such tissue compartmentalization of desensitization mechanisms may have important therapeutic implications with regards to the pharmacological selectivity of putative βARK antagonists (46).

Very similar conclusions were reached in a study by Lohse et al. (59,60) who have examined the effect of inhibition of phosphorylation on the process of rapid desensitization in human epidermal carcinoma A431 cells. Using digitonin to permeabilize the A431 cells, inhibition of protein kinase A was accomplished by exposing the cells to PKI, a heat stable inhibitor peptide while βARK was selectively inhibited by heparin. When desensitization was triggered with low concentrations of agonist only inhibition of the PKA phosphorylation process impaired desensitization whereas at high desensitizing agonist concentrations, inhibition of both PKA and βARK processes reduced desensitization of the system. Thus, these results are also consistent with the hypothesis that phosphorylation of the receptor by PKA might be more important where the system responds to low agonist concentrations whereas both may function in situations where the receptor is more fully occupied. A schematic representation of the two phosphorylation mechanisms involved in the rapid uncoupling of the receptor function are presented in Fig. 2.

On the other hand, very little is known about the structural determinants of the receptor responsible for receptor sequestration or for that matter the exact role it plays in the overall process of desensitization. Sequestration is mildly detectable within a few minutes after agonist exposure. In the process the receptors become removed from the normal membrane environment. This receptor-laiden compartment is defined by two criterion: it is not accessible to hydrophilic ligands and it has a lower density than plasma membranes, allowing for separation of the sequestered compartment by sucrose density gradient centrifugation (45).

Sequestration usually rapidly follows the initial functional uncoupling of receptor from their G-proteins and may play a recycling role in the dephosphorylation mechanism. Sequestered membrane compartment contain phosphatase activity which is capable of rapidly dephosphorylating phosphorylated β-adrenergic receptors (47). Receptors in this compartment can presumably recycle to the plasma membrane where they reacquire full functionality. The mechanism of these processes are not well understood. Cheung et al. (61) have postulated that regions in the third cytoplasmic loop of the β2-adrenergic receptor that are implicated in mediating G-protein coupling are also involved in receptor sequestration. However, recent studies in our laboratory do not support this hypothesis. Campbell et al. (62) have demonstrated that a variety of receptor mutations that are incapable of coupling to $G_s$, sequester normally.

The contribution of receptor sequestration to the overall process of desensitization is also uncertain. Lohse et al. (60) have used concanavalin A treatment of cells which inhibits receptor sequestration to examine this

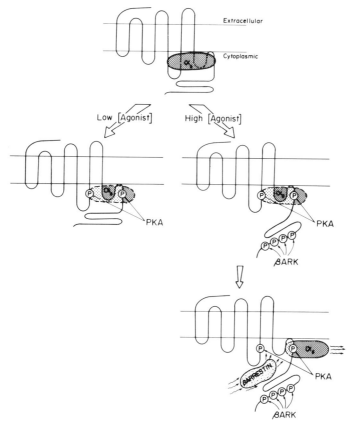

Fig. 2    Schematic representation of the proposed mechanisms of rapid agonist induced desensitization. as represent the a subunit of $G_s$, the stimulatory GTP binding protein.    PKA=protein kinase A.    βARK: β-adrenergic receptor kinase.    P=represent putative sites of phosphorylation on the $\beta_2$-adrenergic receptor for PKA or βARK.    (Reproduced from [46] with permission.)

question.    Treatment of premeabilized A431 cells with ConA did not reduce desensitization unless the two phosphorylation mechanisms alluded to above were blocked.    Under these conditions the extent of desensitization which remains (20-30% of the response) is blockable by ConA (60).    Thus, receptor sequestration would not appear to play a crucial role in the process of rapid agonist mediated desensitization.    Several caveats may apply though.    A431 cells contain a large complement of β-adrenergic receptors (400-500 fmol/mg).    In cells containing much lower concentration of receptors, sequestration may play a more important role.    Direct measurements of receptor sequestration usually indicate that 25-30% of receptors go into this compartment.    Whether the cell contain few or many receptors (spare receptors) may determine what role sequestration plays. Second, sequestration may be more important at longer desensitization times.

The structural basis for receptor down-regulation is not well understood but is known to itself implicate several distinct mechanisms. Apparently, there are both cAMP-dependent and cAMP-independent mechanisms that lead to down regulation (63). The molecular mechanisms involved in down regulation operate over a more prolonged time (hours), which serve to diminish cellular responsiveness by decreasing the total number of cellular receptors. Down regulation may proceed from alterations in the rates of receptor degradation, assembly or processing, messenger RNA (mRNA) stability, and receptor gene transcription (64). Receptor mutants that cannot be phosphorylated by PKA show a slower rate of cAMP-induced receptor down-regulation (63). Down regulation is a complex mechanism and more studies of this phenomenon are necessary to understand its basis.

The adrenergic receptors are part of the family of G protein coupled receptors and have served as prototypes for the study of signal transduction. Many advances in understanding the structural basis for the various functions of adrenergic receptors, and in effect G proteins mediated functions, have occurred over the past decade. The cloning of cDNAs and/or genes for all the major known as well as novel adrenergic receptor subtypes and several G proteins has contributed to this understanding. The insights that are gained into the actions of catecholamines and how various drugs may interact with adrenergic receptors is widely applicable to signal transduction in general. The adrenergic receptor model will undoubtedly prove vital in our under-standing of various pathophysiological situations and hopefully in the future management of these diseases.

## REFERENCES

1.    Gilman AG.  G-Proteins: Transducers of receptors generated signals. *Ann Rev Biochem* 56: 615, 1987.
2.    Lefkowitz RJ, and Caron MG.  Adrenergic Receptors: Models for the study of receptors coupled to guanine nucleotide regulatory proteins. *J Biol Chem* 263: 4993, 1988.
3.    Emorine LJ, Marullo S, Briend-Sutren M-M, Patey G, Tate K, Delavier-Klutchko C, Strosberg AD.  Molecular characterization of the human $\beta_3$-adrenergic receptor. *Science* 245: 1118, 1989.
4.    Bylund DB.  Subtypes of $\alpha_2$-adrenoceptor: Pharmacological and molecular biological evidence coverge. *Trends Pharmacol Sci* 9: 356, 1988.
5.    McGrath J, Wilson V.  $\alpha$-Adrenoceptor subclassification by classical and response-related methods: Same questions, different answers. *Trends Pharmacol Sci* 9: 162, 1988.
6.    Schwinn DA, Lomasney JW, Lorenz W, Szklut P, Fremeau RT, Jr, Yang-Feng TL, Caron MG, Lefkowitz RJ, and Cotecchia S.  Molecular cloning and expression of the cDNA for a novel $\alpha_1$-adrenergic receptor subtype *J Biol Chem* 265: 8183, 1990.
7.    Lomasney JW, Lorenz W, Allen LF, King K, Regan JW, Yang-Feng TL, Caron MG, and Lefkowitz RJ.  Expansion of the $\alpha_2$-adrenergic receptor family: cloning and characterization of a human $\alpha_2$-adrenergic receptor subtype, the gene for which is located on chromosome 2. *Proc Natl Acad Sci* (USA) 87: 5094, 1990.
8.    Sibley DR, Benovic JL, Caron MG, and Lefkowitz RJ.  Regulation of transmembrane signalling by receptor phosphorylation. *Cell* 48: 913, 1987.
9.    Berridge MJ, and Irvine R.  Inositol triphosphate: A novel second messenger in cellular signal transduction. *Nature* 312: 315, 1984.

10.    Yatani A, Codina J, Brown AM, and Birnbaumer L.    Direct activation of mammalian atrial muscarinic K channels by a human erythrocyte pertussis toxin-sensitive G protein, $G_k$. *Science* 235: 207, 1987.

11.    Yatani A, Codina J, Imoto Y, Reeves JP, Birnbaumer L, and Brown AM.    A G protein directly regulates mammalian cardiac calcium channels. *Science* 238: 1288, 1987.

12.    Cotecchia S, Kobilka B, Daniel K, Nolan RD, Lapetina EY, Caron MG, and Lefkowitz RJ.    Multiple second messenger pathways of α-adrenergic receptor subtypes expressed in eukaryotic cells,. *J Biol Chem* 265: 63, 1990.

13.    Peralta EG, Ashkenazi A, Winslow JW, Ramachandran J, and Capon DJ.    Differential regulation of phosphatidyl inositol hydrolysis and adenylyl cyclase by muscarinic receptor subtypes. *Nature* 334: 434, 1988.

14.    Fargin A, Raymond JR, Regan JW, Cotecchia S, Lefkowitz RJ, and Caron MG.    Effector coupling mechanisms of the 5-HT1A receptor, *J Biol Chem* 264: 14848, 1989.

15.    Nathans J, Thomas D, and Hogness DS.    Molecular genetics of human color vision: The genes encoding blue, green, and red pigments. *Science* 232: 193, 1986.

16.    Jones DT, and Reed RR.    Golf: An olfactory neuron specific-G protein involved in odorant signal transduction. *Science* 244: 790, 1989.

17.    Lefkowitz RJ, and Caron MG.    Molecular and regulatory properties of adrenergic receptors. *Recent Prog Horm Res* 43: 469, 1987.

18.    Henderson R, and Unwin PNT.    Three dimensional model of purple membrane obtained by electon microscopy. *Nature* 257: 28, 1975.

19.    Dohlman HG, Bouvier M, Benovic JL, Caron MG, and Lefkowitz RJ. The multiple membrane spanning topography of the β2-adrenergic receptor. Localization of the sites of binding, glycosylation and regulatory phosphorylation by limited proteolysis. *J Biol Chem* 262: 14282, 1987.

20.    Wang H, Lipfert L, Malbon CC, and Bahouth S.    Site-directed anti-peptide antibodies define the topography of the beta-adrenergic receptor. *J Biol Chem* 264: 14424, 1989.

21.    Dixon RAF, Sigal IS, Candelore MR, Register RB, Scattergood W, Rands E, and Strader CD.    Structural features required for ligand binding to the β-adrenergic receptor. *EMBO J* 6: 3269,1987.

22.    Rands E, Candelore MR, Cheung AH, Hill WS, Strader CD, and Dixon RAF.    Mutational analysis of β-adrenergic receptor glycosylation. *J Biol Chem* 265: 10759, 1990.

23.    Fraser CM.    Site-directed mutagenesis of β-adrenergic receptors. Identification of conserved cysteine residues that independently affect ligand binding and receptor activation. *J Biol Chem* 264: 9266, 1989.

24.    Dohlman HG, Caron MG, DeBlasi A, Frielle T, and Lefkowitz RJ.    A role of extracellular disulfide bonded cysteines in the ligand binding function of the β2-adrenergic receptor. *Biochemistry* 29: 2335, 1990.

25.    Karnik SS, Sakmar TP, Chen H-B, and Khorana HG.    Cysteine residues 110 and 187 are essential for the formation of correct structure in bovine rhodopsin. *Proc Natl Acad Sci* (USA) 85: 8459, 1988.

26.    Dohlman HG, Caron MG, Strader CD, Amlaik N, and Lefkowitz RJ. Identification and sequence of a binding site peptide of the β2-adrenergic receptor. *Biochem* 27: 1813, 1988.

27.    Matsu H, Lefkowitz RJ, Caron MG, and Regan JW.    Localization of the fourth membrane spanning domain as a ligand binding site in the human platelet α2-adrenergic receptor. *Biochem* 28: 4125, 1989.

28.　Cirtis CAM, Wheatley M, Bansal S, Birdsall NJM, Eveleight P, Pedder EK, Poyner D, and Hulme EC. Propylbenzilcholine mustard labels an acidic residue in transmembrane helix 3 of the muscarinic receptor. *J Biol Chem* 264: 89,1989.

29.　Wong SK-F, Slaughter C, Ruoho AE, and Ross EM. The catecholamine binding site of the β-adrenergic receptor is formed by juxtaposed membrane-spanning domains. *J Biol Chem* 263: 7925, 1988.

30.　Barclay DC, and Findlay JBC. Labeling of the cytoplasmic domains of ovine rhodopsin with hydrophilic photoaffinity probes. *Biochem J* 220: 75, 1984.

31.　Thomas DD, and Stryer L. Transverse location of the retinal chromophore of rhodopsin in rod outer segment disc membranes. *J Mol Biol* 154: 145, 1982.

32.　Dixon RAF, Sigal IS, Rands E, Register RB, Candelore MR, Blake AD, and Strader CD. Ligand binding to β-adrenergic receptor involves its rhodopsin-like core. *Nature* 326: 73, 1987.

33.　Kobilka BK, MacGregor C, Daniel K, Kobilka TS, Caron MG, and Lefkowitz RJ. Functional activity and regulation of human $\beta_2$-adrenergic receptors expressed in xenopus oocytes. *J Biol Chem* 262: 7321, 1987.

34.　Strader CD, Sigal IS, Candelore MR, Rands E, and Dixon RAF. Identification of residues required for ligand binding to the β-adrenergic receptor. *Proc Natl Acad Sci (USA)* 84: 4384, 1987.

35.　Strader CD, Sigal IS, Candelore MR, Rands E, Hill WS, and Dixon RAF. Conserved aspartic acid residues 79 and 113 of the β-adrenergic receptor have different roles in receptor function. *J Biol Chem* 263: 4052, 1988.

36.　Chung F-Z, Wang C-D Potter PC, Venter JC, and Fraser CM. Site-directed mutagenesis and continuous expression of human β-adrenergic receptors. *J Biol Chem* 263: 4052, 1988.

37.　Frielle T, Daniel K, Caron MG, and Lefkowitz RJ. Structural basis of β-adrenergic receptor subtype specificity studied with chimeric $\beta_1,\beta_2$-adrenergic receptors. *Proc Natl Acad Sc (USA)* 85: 9494, 1988.

38.　Kobilka BK, Kobilka TS, Daniel K, Regan JW, Caron MG, and Lefkowitz RJ. Chimeric $\alpha_2,\beta_2$-adrenergic receptors: Delineation of domains involved in effector coupling and ligand binding specificity. *Science* 240: 1310, 1988.

39.　Strader CD, Candelore MR, Hill WS, Sigal IS, and Dixon RAF. Identification of two serine residues involved in agonist activation of the β-adrenergic receptor. *J Biol Chem* 264: 13572, 1989.

40.　Strader CD, Sigal IS, and Dixon RAF. Structural basis of β-adrenergic receptor function. *FASEB J* 3: 1825, 1989.

41.　O'Dowd BF, Hnatowich M, Regan JW, Leader W.M, Caron MG, and Lefkowitz RJ. Site-directed mutagenesis of the cytoplasmic domains of the human $\beta_2$-adrenergic receptor. *J Biol Chem* 263: 15985, 1988.

42.　O'Dowd BF, Hnatowich M, Caron, MG, Lefkowitz RJ, Bouvier M. Palmitoylation of the human $\beta_2$-adrenergic receptor. J Biol Chem 264: 7564, 1989.

43.　Ovchinnikov YA, Adulaev NG, and Bogachuk AS. Two adjacent cysteine residues in the C-terminal cytoplasmic fragment of bovine rhodopsin are palmitylated. *FEBS Lett* 230:1, 1988.

44.　Stiles GL. Drug and hormonal regulation of the β-adrenergic receptor - adenylate cyclase systems. In *The β-Adrenergic Receptor* (Perkins, J.P., Ed) Humana Press, New York, pp.345-386, 1991.

45.     Perkins JP, Hausdorff WP, and Lefkowitz RJ.   Mechanisms of ligand-induced desensitization of β-adrenergic receptors. In *The β-Adrenergic Receptor* (Perkins, J.P., Ed) Humana Press, New York, pp. 73-179, 1991.

46.     Hausdorff WP, Caron MG, and Lefkowitz RJ. Turning off the signal: desensitization of the β-adrenergic receptor function. *FASEB J* 4: 2881, 1990.

47.     Sibley DR, Benovic JL, Caron MG, and Lefkowitz .J. Regulation of transmembrane signaling by receptor phosphorylation *Cell* 48: 913, 1987.

48.     Benovic JL, Pike LJ, Cerion RA, Staniszewski C, Yoshimasa T, Codina J, Caron MG, and Lefkowitz RJ.   Phosphorylation of the mammalian β-adrenergic receptor by cyclic AMP-dependent kinase. *J Biol Chem* 260: 7094, 1985.

49.     Bouvier M, Leeb-Lundberg LMF, Benovic JL, Caron MG, and Lefkowitz RJ.   Regulation of adrenergic receptor function by phosphorylation. *J Biol Chem* 262: 3106, 1987.

50.     Benovic JL, Strasser RH, Caron MG, and Lefkowitz RJ.   β-Adrenergic receptor kinase: Identification of a novel protein kinase that phosphorylates the agonist-occupied form of the receptor. *Proc Natl Acad Sci (USA)* 83: 2797, 1986.

51.     Benovic JL, Mayor F, Jr, Staniszewski C. Lefkowitz RJ, and Caron MG.   Purification and characterization of the β-adrenergic receptor kinase. *J Biol Chem* 262: 9026, 1987.

52.     Benovic JL, DeBlasi A, Stone WC, Caron MG, and Lefkowitz RJ.   β-Adrenergic receptor kinase: Primary structure delineates a multigene family. *Science* 246:2 35, 1989.

53.     Benovic JL, Kuhn H, Weyand I, Codina J, Caron MG, and Lefkowitz RJ. Functional desensitization of the isolated β-adrenergic receptor by the β-adrenergic receptor kinase: Potential role of an analog of the retinal binding protein arrestin (48-kDa). *Proc Natl Acad Sci (USA)* 84: 8879, 1987.

54.     Wilden U, Hall SW, and Ku Ohn H. Phosphodiesterase activation by photoexited rhodopsin is quenched when rhodopsin is phosphorylated and binds to the intrinsic 48-kDa protein of the rod outer segment. *Proc Natl Acad Sci (USA)* 83: 1174, 1986.

55.     Shinohara T, Dietzschold B, Craft CM, Wistow G, Early JJ, Donoso LA, Horowitz J, and Tao R.   Primary and secondary structure of bovine retinal S-antigen (48-kDa protein). *Proc Natl Acad Sci (USA)* 84: 6975, 1987.

56.     Lohse MJ, Benovic JL, Codina J, Caron MG, and Lefkowitz RJ. β-Arrestin: A protein that regulates β-adrenergic receptor function. *Science* 248: 1547, 1990.

57.     Benovic JL, Mayor F, Jr, Somers RL, Caron MG, and Lefkowitz RJ. Light-dependent phosphorylation of rhodopsin by β-adrenergic receptor kinase. *Nature* 322: 867, 1986.

58.     Hausdorff WP, Bouvier M, O'Dowd BF, Irons GP, Caron MG, and Lefkowitz RJ.   Phosphorylation sites on two domains of the $β_2$-adrenergic receptor are involved in distinct pathways of receptor desensitization. *J Biol Chem* 264: 12657, 1989.

59.     Lohse MJ, Lefkowitz RJ, Caron MG, and Benovic JL. Inhibition of β-adrenergic receptor kinase prevents rapid homologous desensitization of $β_2$-adrenergic receptors. *Proc Natl Acad Sci (USA)* 86: 3011, 1989.

60.     Lohse MJ, Benovic JL, Caron MG, and Lefkowitz RJ.   Multiple pathways of rapid $β_2$-adrenergic receptor desensitization: Delineation with specific inhibitors. *J Biol Chem* 265: 3202, 1990.

61.    Cheung AH, Sigal IS, Dixon RAF.,and Strader CA.    Agonist-promoted sequestration of the $\beta_2$-adrenergic receptor requires regions involved in functional coupling with $G_s$. *Mol Pharmacol* 34: 132, 1989.

62.    Campbell PT, Hnatowich M, O'Dowd BF, Caron MG, Lefkowitz RJ. and Hausdorff, W.P., Mutations of the human $\beta_2$-adrenergic receptor that impair coupling to $G_s$ interfere with receptor down-regulation but not sequestration. *Mol Pharmacol* 39: 192, 1991.

63.    Bouvier M, Collins S, Caron MG, and Lefkowitz RJ.    Two distinct pathways for cAMP mediated down regulation of the $\beta_2$-adrenergic receptor: Phosphorylation of the receptor and regulation of its mRNA level. *J Biol Chem* 264: 16786, 1989.

64.    Collins S, Bouvier M, Bolanowski MA, Caron MG, and Lefkowitz RJ. cAMP stimulates transcription of $\beta_2$-adrenergic receptor gene in response to short-term agonist exposure. *Proc Natl Acad Sci (USA)* 86: 4853, 1989.

# CONTRIBUTORS

**H. S. Ahn**, Department of Pharmacology, Schering-Plough Research Division, Bloomfield, NJ 07003

**J. J. Bahl**, Departments of Internal Medicine, Physiology and Pharmacology, University Heart Center, Arizona Health Sciences Center, Tucson, AZ 85724

**Robert L. Barchi**, Mahoney Institute of Neurological Sciences, University of Pennsylvania School of Medicine, Philadelphia, PA 19104

**L. Birnbaumer**, Departments of Molecular Physiology and Biophysics and Cell Biology, Baylor College of Medicine, One Baylor Plaza, Houston, Texas 77030

**A. M. Brown**, Departments of Molecular Physiology and Biophysics, Baylor College of Medicine, One Baylor Plaza, Houston, Texas 77030

**R. Allan Buchholz**, Department of Cardiovascular Pharmacology, Sterling Research Group, Rensselaer, NY 12144

**Aram V. Chobanian**, Boston University School of Medicine, Boston, MA 02118

**M. Cable**, Department of Pharmacology, Schering-Plough Research Division, Bloomfield, NJ 07003

**Marc G. Caron**, Howard Hughes Medical Institute, Duke University Medical Center, Durham, North Carolina 27710

**Jacqueline Cilea**, Bockus Research Institute, Graduate Hospital, Philadelphia, PA 19146

**J. Codina**, Department of Cell Biology, Baylor College of Medicine, One Baylor Plaza, Houston, Texas 77030

**Robert H. Cox**, Bockus Research Institute, The Graduate Hospital and Department of Physiology, The University of Pennsylvania, Philadelphia, PA 19104

**Wayne R. Cumiskey**, Department of Cardiovascular Pharmacology, Sterling Research Group, Rensselaer, NY 12144

**Ronald L. Dundore**, Department of Cardiovascular Pharmacology, Sterling Research Group, Rensselaer, NY 12144

**J. G. Edwards**, Departments of Internal Medicine, Physiology and Pharmacology, University Heart Center, Arizona Health Sciences Center, Tucson, AZ 85724

**D. Eliades**, Department of Physiology, Uniformed Services University, Bethesda, MD 20814

**Alan M. Ezrin**, Department of Cardiovascular Pharmacology, Sterling Research Group, Rensselaer, NY 12144

**I. L. Flink**, Departments of Internal Medicine, Physiology and Pharmacology, University Heart Center, Arizona Health Sciences Center, Tucson, AZ 85724

**M. Foster**, Department of Pharmacology, Schering-Plough Research Division, Bloomfield, NJ 07003

**Harry A. Fozzard**, Cardiac Electrophysiology Laboratories, Departments of Medicine and the Pharmacological & Physiological Sciences, and the Committee on Cell Physiology, The University of Chicago, Chicago, IL 60637

**Edward D. Frohlich**, Vice President for Academic Affairs, Alton Ochsner Medical Foundation, New Orleans, LA 70120

**Brinda B. Geisbuhler**, Departments of Physiology and Pharmacology, University of Missouri, Columbia, MO 65212

**F. J. Haddy**, Department of Physiology, Uniformed Services University, Bethesda, MD 20814

**Alex L. Harris**, Department of Cardiovascular Pharmacology, Sterling Research Group, Rensselaer, NY 12144

**Douglas W. P. Hay**, Department of Physiology and Biophysics, Howard University College of Medicine, Washington, DC 20059

**Allan W. Jones**, Departments of Physiology and Pharmacology, University of Missouri, Columbia, MO 65212

**Susan B. Jones**, Departments of Physiology and Pharmacology, University of Missouri, Columbia, MO 65212

**G. Kirsch**, Departments of Molecular Physiology and Biophysics, Baylor College of Medicine, One Baylor Plaza, Houston, Texas 77030

**Edward G. Lakatta**, Laboratory of Cardiovascular Science, Gerontology Research Center, National Institute on Aging, National Institutes of Health, Baltimore, MD 21224

**King C. Lee**, Department of Cardiovascular Pharmacology, Sterling Research Group, Rensselaer, NY 12144

**Robert J. Lefkowitz**, Howard Hughes Medical Institute, Duke University Medical Center, Durham, North Carolina 27710

**Robert S. Moreland**, Bockus Research Institute, Graduate Hospital, Philadelphia, PA 19146

**Suzanne Moreland**, The Bristol-Myers Squibb Pharmaceutical Research Institute, Princeton, NJ 08540

**Eugene Morkin**, Departments of Internal Medicine, Physiology and Pharmacology, University Heart Center, Arizona Health Sciences Center, Tucson, AZ 85724

**Evangeline D. Motley**, Department of Pharmacology, SmithKline Beecham Pharmaceuticals, King of Prussia, PA 19406, Department of Physiology and Biophysics, Howard University College of Medicine, Washington, DC 20059

**Andrew J. Nichols**, Department of Physiology and Biophysics, Howard University College of Medicine, Washington, DC 20059

**Junji Nishimura**, Department of Pharmacology, University of Miami, School of Medicine, Miami, FL 33101

**Gary K. Owens**, Department of Physiology, University of Virginia School of Medicine, Charlottesville, VA 22908

**Edward D. Pagani**, Department of Cardiovascular Pharmacology, Sterling Research Group, Rensselaer, NY 12144

**M. B. Pamnani**, Department of Physiology, Uniformed Services University, Bethesda, MD 20814

**B. J. R. Pitts**, Department of Pharmacology, Schering-Plough Research Division, Bloomfield, NJ 07003

**Russell L. Prewitt**, Department of Physiology, Eastern Virginia Medical School, Norfolk, Virginia 23501

**Neil S. Roth**, Howard Hughes Medical Institute, Duke University Medical Center, Durham, North Carolina 27710

**Robert R. Ruffolo, Jr.**, Department of Pharmacology, SmithKline Beecham Pharmaceuticals, King of Prussia, PA 19406

**Nancy J. Rusch**, Department of Physiology, Medical College of Wisconsin, Milwaukee, WI 53226

**B. Schubert**, Departments of Molecular Physiology and Biophysics, Baylor College of Medicine, One Baylor Plaza, Houston, Texas 77030

**Shivendra D. Shukla**, Departments of Physiology and Pharmacology, University of Missouri, Columbia, MO 65212

**Paul J. Silver**, Department of Cardiovascular Pharmacology, Sterling Research Group, Rensselaer, NY 12144

**Jacquelyn M. Smith**, Department of Physiology, Chicago College of Osteopathic Medicine, Downers Grove, IL 60515

**William J. Stekiel**, Department of Physiology, Medical College of Wisconsin, Milwaukee, WI 53226

**B. T. Swindall**, Department of Physiology, Uniformed Services University, Bethesda, MD 20814

**E. J. Sybertz**, Department of Pharmacology, Schering-Plough Research Division, Bloomfield, NJ 07003

**R. W. Tsika**, Departments of Internal Medicine, Physiology and Pharmacology, University Heart Center, Arizona Health Sciences Center, Tucson, AZ 85724

**Cornelis van Breemen**, Department of Pharmacology, University of Miami, School of Medicine, Miami, FL 33101

**A. M. J. VanDongen**, Departments of Molecular Physiology and Biophysics, Baylor College of Medicine, One Baylor Plaza, Houston, Texas 77030

**Duo H. Wang**, Department of Physiology, Eastern Virginia Medical School, Norfolk, Virginia 23501

**A. Yatani**, Departments of Molecular Physiology and Biophysics, Baylor College of Medicine, One Baylor Plaza, Houston, Texas 77030

# INDEX